MATHEMATICS RESEARCH DEVELOPMENTS

AN INTRODUCTION TO THE EXTENDED KALMAN FILTER

MATHEMATICS RESEARCH DEVELOPMENTS

Additional books and e-books in this series can be found on Nova's website under the Series tab.

MATHEMATICS RSEARCH DEVELOPMENTS

AN INTRODUCTION TO THE EXTENDED KALMAN FILTER

MATTHIAS HOLLAND
EDITOR

Copyright © 2020 by Nova Science Publishers, Inc.

All rights reserved. No part of this book may be reproduced, stored in a retrieval system or transmitted in any form or by any means: electronic, electrostatic, magnetic, tape, mechanical photocopying, recording or otherwise without the written permission of the Publisher.

We have partnered with Copyright Clearance Center to make it easy for you to obtain permissions to reuse content from this publication. Simply navigate to this publication's page on Nova's website and locate the "Get Permission" button below the title description. This button is linked directly to the title's permission page on copyright.com. Alternatively, you can visit copyright.com and search by title, ISBN, or ISSN.

For further questions about using the service on copyright.com, please contact:
Copyright Clearance Center
Phone: +1-(978) 750-8400   Fax: +1-(978) 750-4470   E-mail: info@copyright.com.

NOTICE TO THE READER

The Publisher has taken reasonable care in the preparation of this book, but makes no expressed or implied warranty of any kind and assumes no responsibility for any errors or omissions. No liability is assumed for incidental or consequential damages in connection with or arising out of information contained in this book. The Publisher shall not be liable for any special, consequential, or exemplary damages resulting, in whole or in part, from the readers' use of, or reliance upon, this material. Any parts of this book based on government reports are so indicated and copyright is claimed for those parts to the extent applicable to compilations of such works.

Independent verification should be sought for any data, advice or recommendations contained in this book. In addition, no responsibility is assumed by the Publisher for any injury and/or damage to persons or property arising from any methods, products, instructions, ideas or otherwise contained in this publication.

This publication is designed to provide accurate and authoritative information with regard to the subject matter covered herein. It is sold with the clear understanding that the Publisher is not engaged in rendering legal or any other professional services. If legal or any other expert assistance is required, the services of a competent person should be sought. FROM A DECLARATION OF PARTICIPANTS JOINTLY ADOPTED BY A COMMITTEE OF THE AMERICAN BAR ASSOCIATION AND A COMMITTEE OF PUBLISHERS.

Additional color graphics may be available in the e-book version of this book.

Library of Congress Cataloging-in-Publication Data

ISBN: 978-1-53618-875-2

Published by Nova Science Publishers, Inc. † New York

CONTENTS

Preface vii

Chapter 1 EKF-Based Orbit Determination Using Single Ground Tracking Antenna Measurements 1
Chingiz Hajiyev and Ahmet Sofyalı

Chapter 2 Kalman Filters for Descriptor Systems Applied in Domain of Fault Tolerant Control: A Review 43
Tigmanshu Patel, M. S. Rao, Jalesh L. Purohit and V. A. Shah

Chapter 3 Convergence of the Discrete Extended Kalman Filter in Nonlinear Deterministic Systems with Noisy Outputs 83
Leonardo Esau Herrera and Jaime Herrera

Index 103

PREFACE

An Introduction to the Extended Kalman Filter first presents a study wherein a two-stage approach for the estimation of a spacecraft's position and velocity using single station antenna tracking data is proposed.

Since the Kalman filter and its variants are widely used for estimation in diverse domains, the authors also present a review of fault detection, diagnosis and fault tolerant control of descriptor/differential algebraic equation systems specifically focused on the Kalman filter and its variants.

The closing contribution provides insight into the intrinsic convergence of the extended Kalman filter when operated in the stochastic frame for the class of systems and outputs considered.

In Chapter 1, a two-stage approach for estimation of spacecraft's position and velocity by single station antenna tracking data is proposed. In the first stage, direct nonlinear antenna measurements are transformed to linear x-y-z coordinate measurements of spacecraft's position, and statistical characteristics of orbit determination errors are analyzed. Variances of orbit parameters' errors are chosen as the accuracy criteria. In the second stage, the outputs of the first stage are improved by the designed EKF for estimation of the spacecraft's position and velocity on indirect linear x-y-z measurements.

Two recursive EKF algorithms are examined: the first designed filter estimates the position and velocity of the spacecraft on direct nonlinear range-azimuth-elevation measurements; the second one carries out the estimation on indirect linear x-y-z position measurements. The examined filters are compared through geostationary satellite orbital motion simulations, and the recommendations on their application for various spacecraft missions are given.

As explained in Chapter 2, safety, reliability, and dependability of complex industrial processes are of significant importance. Mathematical model based process monitoring and fault tolerant control (FTC) can be employed to address issues related to safe and reliable process operation. The fault diagnosis (FD) and FTC of systems described by ordinary differential equations (ODEs) are relatively well addressed in the literature. However, a large class of systems can be conveniently expressed in descriptor/Differential Algebraic Equation (DAE) form. Descriptor systems tend to possess peculiar properties which in turn demand radically different approaches to address its issues as compared to ODEs. Consequently, the problem of FD and FTC of descriptor systems has received critical attention and continues to be investigated. In spite, a systematic review on FD and FTC of descriptor systems is by and large missing in literature. Since Kalman filter and its variants are widely used for estimation in diverse domains, the authors present a review of fault detection, diagnosis and fault tolerant control of descriptor/DAE systems specifically focused on Kalman filter and its variants. Firstly, the domain of FD and FTC is summarized and alternate terminologies are explicitly put forth. Further, the properties and issues of descriptor system are outlined. The approaches for fault detection of descriptor systems are evaluated wherein different Kalman filters are clarified. Sparse work of fault isolation, fault estimation and fault tolerant control of descriptor systems is also outlined. Finally, various examples of descriptor systems across literature are summarized and put forth. The tools and solvers available for descriptor systems are highlighted.

The convergence of the extended Kalman filter in nonlinear discrete deterministic systems corrupted by white Gaussian noise in its outputs is presented in Chapter 3. In the linear counterpart, the convergence was shown for the Kalman filter. To extend the results to the nonlinear frame and consequently cover a larger class of systems, useful tools from the linear counterpart are brought to support the present study. Some assumptions in the linearized structure of the nonlinear systems and its outputs are sufficient to demonstrate, via Lyapunov analysis, that the estimated state provided by the extended Kalman filter converges to that of the examined nonlinear systems. As a deterministic focus is considered, asymptotic convergence is achieved. The proposed contribution thus gives an insight into the intrinsic convergence of the extended Kalman filter when operated in the stochastic frame for the class of systems and outputs considered. As a testbed, bifurcation parameter estimation is addressed in a three-dimensional chaotic system and its output to illustrate the convergence. Numerical results show the asymptotic convergence.

In: An Introduction to the Extended ... ISBN: 978-1-53618-875-2
Editor: M. Holland © 2020 Nova Science Publishers, Inc.

Chapter 1

EKF-BASED ORBIT DETERMINATION USING SINGLE GROUND TRACKING ANTENNA MEASUREMENTS

Chingiz Hajiyev[1,] and Ahmet Sofyalı[2]*

[1]Aeronautical Engineering Department,
Istanbul Technical University, Istanbul, Turkey

[2]Astronautical Engineering Department,
Istanbul Technical University, Istanbul, Turkey

ABSTRACT

In this study, a two-stage approach for estimation of spacecraft's position and velocity by single station antenna tracking data is proposed. In the first stage, direct nonlinear antenna measurements are transformed to linear x-y-z coordinate measurements of spacecraft's position, and statistical characteristics of orbit determination errors are analyzed. Variances of orbit parameters' errors are chosen as the accuracy criteria. In the second stage, the outputs of the first stage are improved by the designed EKF for estimation of the spacecraft's position and velocity on indirect linear x-y-z measurements.

Two recursive EKF algorithms are examined: the first designed filter estimates the position and velocity of the spacecraft on direct nonlinear range-

---

* Corresponding Author's Email: cingiz@itu.edu.tr.

azimuth-elevation measurements; the second one carries out the estimation on indirect linear x-y-z position measurements. The examined filters are compared through geostationary satellite orbital motion simulations, and the recommendations on their application for various spacecraft missions are given.

Keywords: spacecraft, single station antenna, estimation, extended Kalman filter, indirect measurements

INTRODUCTION

Orbit determination is the process of finding the best approximation of spacecraft's position over time using observations of their position or velocity, where their motion is described by imperfect dynamic models.

In orbit determination of spacecraft, the dynamic system and the measurement equations are of nonlinear nature. It is a nonlinear problem, in which the disturbing forces are not easily modeled. The problem consists of estimating variables that completely describe the body's trajectory in space, processing a set of information related to the considered body [1]. A tracking antenna on Earth or sensors such as GPS receivers, magnetometers, etc. can perform such observations.

The work [2] describes an optimal iterative algorithm capable of determining the orbital parameters by using the antenna pointing angles, which are recorded in the tracking of the passing satellite. The algorithm is optimal in the sense that it minimizes the noise effects of the noisy measurements and the numerical uncertainties of the propagation. The method is originated from the Least Squares Estimation (LSE) algorithm, which, by using the theory of the Extended Kalman Filter (EKF), is suitably modified to reduce the disturbances on the estimation error.

Orbit determination accuracy improvement for a geostationary satellite with single station antenna tracking data is investigated in [3]. In this study, an operational orbit determination (OD) system for the geostationary satellite mission requires accurate satellite positioning data to accomplish image navigation registration on the ground. Ranging and tracking data, which is provided by a single ground station, is used to determine the orbit of the geostationary satellite in normal operation. However, the orbital longitude of the geostationary satellite is so close to those of satellite tracking sites that emerging geometric singularity affects observability [4]. Applying an azimuth bias estimation using the ranging and tracking data provided by two stations is a method to solve for the azimuth bias of a single station in singularity. Using only single-station data with the correction of the azimuth bias, OD succeeds to achieve three-sigma position accuracy of the order of 1.5 km root-sum square.

Localization of spacecraft is usually very accurate when GPS measurements are available [5]. The problem becomes more challenging when GPS signals are not available, for instance in high-Earth orbits or in long range missions such as Earth-to-Moon transfers. In these cases, spacecraft navigation is often handled by ground-based tracking stations, thus making it unfeasible for low-cost spacecraft missions. In order to make spacecraft fully autonomous, it is necessary to devise self-localization and navigation algorithms relying on measurements provided by onboard sensors. In [6], the problem of spacecraft self-localization is addressed using angular measurements. A dynamic model of the spacecraft accounting for several perturbing effects, such as Earth and Moon gravitational field asymmetry and errors associated with the Moon ephemerides, is employed. It is assumed that the navigation system is able to estimate the spacecraft's attitude (by using a star tracker sensor), and the spacecraft is equipped with line-of-sight sensors providing measurements of elevation and azimuth of the Moon and the Earth with respect to the spacecraft reference system. Range measurements, which are often difficult to obtain or are not sufficiently reliable, are not required.

In [7], a comparison of Extended Kalman Filter (EKF) and Unscented Kalman Filter (UKF) for spacecraft localization via angle measurements is performed. In the study, performances of two nonlinear estimators are compared for the localization of a spacecraft. It is assumed that range measurements are not available, and the localization problem is tackled on the basis of angle-only measurements. The dynamic model of the spacecraft is the same as in [6]. The measurement process is based on elevation and azimuth of Moon and Earth with respect to the spacecraft reference system. The position and velocity of the spacecraft are estimated using both EKF and UKF. The behaviors of the filters are compared on two sample missions: Earth-to-Moon transfer and geostationary orbit raising.

Orbit determination techniques in [8] are used to estimate the position and velocity of a debris object in orbit using range, azimuth, and elevation measurements obtained from Space Surveillance Network (SSN) sensors. The continuous-discrete extended Kalman filter is used to estimate the debris' orbit.

In [9], the non-recursive batch filter has been presented and utilized for satellite orbit determination. Using the unscented transformation, a non-recursive batch filter is developed without any traditional linearization process. For the orbit determination system, the range, azimuth, and elevation angles of the satellite measured by ground tracking stations are used as observations. For evaluation and verification of the presented batch filter's performance, the results are compared with those of the batch least squares filter for various initial errors in position and velocity, measurement sampling periods, and measurement errors. For relatively small initial errors or short measurement sampling periods or small measurement errors, the accuracy of the orbit determination is similar in both filters. Under large initial errors or long measurement sampling periods or large measurement errors, the presented non-recursive batch filter demonstrates more robust and stable convergence than the existing batch least squares filter's.

Recently, the so-called Gaussian-mixture Kalman filters had been revisited by Mark. L. Psiaki. After a line of works by Psiaki and colleagues, he applied that method to the satellite trajectory estimation problem [10]. The introduced orbit determination Kalman filter has been developed especially to perform by using sparsely available short arcs of angles-only data such as from 20 seconds to 5 minutes per night for near geosynchronous orbits.

Most recently, the orbit determination problem of satellites performing unknown continuous maneuvers has been dealt with by a proposed augmentation to the so-called unbiased minimum-variance input and state estimation method [11]. The augmentation is the estimation of maneuver acceleration by innovations of previous N steps in the recursive process by assuming maneuvers to be non-agile. Range and range-rate measurements are accepted to be available by three ground radar stations. Simulations demonstrated significant improvement in the acceleration estimation accuracy compared to the baseline method's accuracy.

In general, for the orbit determination purpose, Kalman filtering technique is used. Antenna tracking data can be processed by Kalman filter in various methods [1-15]. Because the antenna measurements (azimuth, elevation, and distance) are non-linear with respect to the state variables, the process of location estimation of spacecraft by using antenna tracking data is non-linear and can only be solved by EKF or UKF.

In this study, two-stage estimation of spacecraft's position and velocity by single station antenna tracking data is proposed. In the first stage, the direct nonlinear antenna measurements are transformed to the linear x-y-z coordinate measurements of spacecraft's position, and their errors' variances are evaluated. In the second stage, the outputs of the first stage are improved by EKF for estimation of the spacecraft's position and velocity based on indirect linear x-y-z measurements.

Two recursive EKF algorithms are examined: the first conventional filter estimates spacecraft's position and velocity on the direct nonlinear range-azimuth-elevation measurements; the second one estimates the spacecraft position and velocity on indirect linear x-y-z measurements. The results of examined filters are compared through geostationary satellite orbital motion simulations. Simulations prove that the proposed two-stage estimation method straightened the single station antenna tracking data very well and gives better results than the existing method.

EKF FOR ESTIMATION OF SPACECRAFT POSITION AND VELOCITY USING DIRECT RANGE-AZIMUTH-ELEVATION MEASUREMENTS

The Mathematical Model of the Spacecraft Orbital Motion

Kepler equations system is one of the systems that can define elliptic orbits of spacecraft. This equations system includes three equations in differential form [16]:

$$\frac{d^2x}{dt^2} = -\gamma M \frac{x}{r^3} \quad (1)$$

$$\frac{d^2y}{dt^2} = -\gamma M \frac{y}{r^3} \quad (2)$$

$$\frac{d^2z}{dt^2} = -\gamma M \frac{z}{r^3} \quad (3)$$

where $\gamma = 6.67 \times 10^{-11} m^3/kgs^2$ is the Kepler constant, x, y, and z are the Descartes coordinates of the spacecraft in Earth-Centered-Inertial (ECI) coordinate frame, $r = \sqrt{x^2 + y^2 + z^2}$ is the range between spacecraft's center of mass and Earth's center of mass, $M = 5.976 \times 10^{24} kg$ is the mass of the Earth. Write Equations (1)-(3) in the form of six difference equations as;

$$u_{i+1} = u_i - \Delta t\, \gamma M \frac{x_i}{r_i^3} + w_u \quad (4)$$

$$v_{i+1} = v_i - \Delta t\, \gamma M \frac{y_i}{r_i^3} + w_v \quad (5)$$

$$w_{i+1} = w_i - \Delta t\, \gamma M \frac{z_i}{r_i^3} + w_w \quad (6)$$

$$x_{i+1} = x_i + \Delta t u_i + w_x \quad (7)$$

$$y_{i+1} = x_i + \Delta t v_i + w_y \quad (8)$$

$$z_{i+1} = x_i + \Delta t w_i + w_z \quad (9)$$

In the equations; Δt is the sampling time; u, v, and w are the velocities in x, y, and z directions, respectively; $w_{(.)}$ is the Gaussian white noise with zero mean. As a result, it is possible to propagate the orbital position of the spacecraft for a desired time period by using these six equations.

By means of a ground tracking antenna, the distance from the antenna to the spacecraft (ρ), and the azimuth (α) and elevation (β) angles of the spacecraft are measured. The measurement equations are expressed as follows:

$$\rho = \sqrt{\rho_U^2 + \rho_E^2 + \rho_N^2} \qquad (10)$$

$$\alpha = \tan^{-1} \frac{\rho_E}{\rho_N} \qquad (11)$$

$$\beta = \sin^{-1} \frac{\rho_U}{\rho} \qquad (12)$$

Here, ρ_U, ρ_E, and ρ_N are the components of ρ in the Up (Zenith)-East-North (UEN) coordinate system.

Estimation of Spacecraft Position and Velocity on Nonlinear Antenna Measurements

Consider the mathematical model of spacecraft's orbital motion in matrix form as;

$$X(k+1) = f(X(k)) + GW(k) \qquad (13)$$

where $X(k)$ is the state vector, $W(k)$ is the system noise, G is the transition matrix of the system noise.

The state vector of the orbital motion of spacecraft can be written in the following form;

$$X = \begin{bmatrix} x & u & y & v & z & w \end{bmatrix}^T,$$

The non-linear measurement equation is

$$Z(k) = h(X(k), k) + V(k) \tag{14}$$

where $Z(k)$ is the measurement vector, $h(X(k), k)$ is a nonlinear measurement model mapping current state to measurements, $V(k)$ is the random measurement noise.

It is assumed that both noise vectors $W(k)$ and $V(k)$ are white Gaussian. Their mean values and covariances are given below;

$$E[W(k)] = 0; E[V(k)] = 0;$$
$$E[W(k)W^T(j)] = Q(k)\delta(kj);$$
$$E[V(k)V^T(j)] = R(k)\delta(kj).$$

where E is the statistical averaging operator, $\delta(kj)$ is the Kronecker symbol

$$\delta(kj) = \begin{cases} 1, k = j; \\ 0, k \neq j. \end{cases}$$

Equation (14), describing the observations made, can be expressed as a Taylor series about the state prediction $\hat{X}(k/k-1)$ [17]

$$Z(k) = h(X(k)) + V(k) = h(\hat{X}(k/k-1)$$
$$+ \nabla h_X(k)\left[\hat{X}(k/k-1) - X(k)\right] \tag{15}$$
$$+ O\left[\left|\hat{X}(k/k-1) - X(k)\right|^2\right] + V(k)$$

where $\nabla h_X(k)$ is the Jakobian of h evaluated at $X(k) = \hat{X}(k/k-1)$. Ignoring second- and higher-order terms and taking expectations conditioned on the first (k-1) observations give an equation for the predicted observation

$$E\left[Z(k)/Z^{*(k-1)}\right] \approx E\begin{bmatrix} h\ \hat{X}(k/k-1) \\ +\nabla h_X(k)\begin{bmatrix}\hat{X}(k/k-1)\\-X(k)\end{bmatrix} \\ +V(k)/Z^{*(k-1)} \end{bmatrix} \quad (16)$$

$$= h\ \hat{X}(k/k-1)$$

Here, $Z^{*(k-1)} = \{Z(1), Z(2), ..., Z(k-1)\}$.

Equation (16) depends on the fact that the state prediction error and the observation noise both have zero mean. After taking an observation $Z(k)$, the innovation can be found by subtracting the predicted observation as

$$\Delta(k) = Z(k) - h\ \hat{X}(k/k-1) \ . \quad (17)$$

The estimation of the state vector $\hat{X}(k/k)$ and the correlation matrix of error $P(k/k)$ are found by EKF of the following form:

The state estimation

$$\hat{X}(k+1/k+1) = \hat{X}(k+1/k) + K(k+1)\Delta(k+1) \quad (18)$$

the innovation

$$\Delta(k+1) = Z(k+1) - h\ \hat{X}(k+1/k) \quad (19)$$

the state prediction

$$\hat{X}(k+1/k) = f(\hat{X}(k/k)) \quad (20)$$

the Kalman gain

$$K(k+1) = P(k+1/k)\nabla h_X^T(k+1)\begin{bmatrix} \nabla h_X(k+1)P(k+1/k)\nabla h_X^T(k+1) \\ +R(k+1) \end{bmatrix}^{-1} \quad (21)$$

the covariance estimation

$$P(k+1/k+1) = \begin{bmatrix} I - K(k+1)\nabla h_X(k+1) \end{bmatrix} P(k+1/k) \quad (22)$$

the covariance prediction

$$P(k+1/k) = F_X P(k/k) F_X^T + GQ(k)G^T \quad (23)$$

$F_X = \left[\dfrac{\partial F(X)}{\partial X} \right]_{\hat{X}(k)}$ is the Jacobian matrix of the function $F(X)$ evaluated at $X(k) = \hat{X}(k/k)$. The expressions for the matrices F_X and ∇h_X can be written in the following forms

$$F_X = \begin{bmatrix} \partial x/\partial x & \partial x/\partial u & \partial x/\partial y & \partial x/\partial v & \partial x/\partial z & \partial x/\partial w \\ \partial u/\partial x & \partial u/\partial u & \partial u/\partial y & \partial u/\partial v & \partial u/\partial z & \partial u/\partial w \\ \partial y/\partial x & \partial y/\partial u & \partial y/\partial y & \partial y/\partial v & \partial y/\partial z & \partial y/\partial w \\ \partial v/\partial x & \partial v/\partial u & \partial v/\partial y & \partial v/\partial v & \partial v/\partial z & \partial v/\partial w \\ \partial z/\partial x & \partial z/\partial u & \partial z/\partial y & \partial z/\partial v & \partial z/\partial z & \partial z/\partial w \\ \partial w/\partial x & \partial w/\partial u & \partial w/\partial y & \partial w/\partial v & \partial w/\partial z & \partial w/\partial w \end{bmatrix}; \quad (24)$$

$$\nabla h_X = \begin{bmatrix} \partial \rho/\partial x & \partial \rho/\partial u & \partial \rho/\partial y & \partial \rho/\partial v & \partial \rho/\partial z & \partial \rho/\partial w \\ \partial \alpha/\partial x & \partial \alpha/\partial u & \partial \alpha/\partial y & \partial \alpha/\partial v & \partial \alpha/\partial z & \partial \alpha/\partial w \\ \partial \beta/\partial x & \partial \beta/\partial u & \partial \beta/\partial y & \partial \beta/\partial v & \partial \beta/\partial z & \partial \beta/\partial w \end{bmatrix} \quad (25)$$

EKF FOR ESTIMATION OF SATELLITE POSITION AND VELOCITY BASED ON INDIRECT POSITION MEASUREMENTS

As seen from Equations (18)-(25), both the equations of system and measurements are nonlinear in EKF for estimation of spacecraft's position and velocity using direct range-azimuth-elevation measurements. EKF is a highly preferred method for orbit estimation of spacecraft [18]. However, EKF in Equations (18)-(25) has some disadvantages, especially for the highly nonlinear systems. Generally, this is caused by the mandatory linearization phase of the EKF procedure, so, by the derived Jacobian matrices in Equations (24) and (25).

Inputting the position components of the spacecraft in ECI reference frame calculated from the real range measurements in UEN reference frame to the filter leads to a linear measurement function. The conventional approach of inputting the real range-azimuth-elevation measurements directly to the filter results in a highly nonlinear measurement function. The corresponding Jacobian H matrix consists of complex terms, which increases the computational load significantly. For most of the applications, generation of Jacobians is hard, time consuming, and prone to human errors [19]. Nonetheless, linearization brings about an unstable filter performance when the time step intervals for update are not sufficiently small and that results in filter divergence [20]. Per contra, small time step intervals increase the computational burden because of the larger number of Jacobian calculations.

Below, two-stage estimation of spacecraft's position and velocity by single station antenna tracking data, where the measurement nonlinearities are eliminated, is proposed. In first stage, the direct nonlinear antenna measurements are transformed to the linear x-y-z coordinate measurements of spacecraft's position, and their errors' variances are evaluated. In the second stage, the solutions of the first stage are improved by EKF for estimation of the spacecraft position and velocity on indirect linear x-y-z measurements. The simulation results show that the proposed two-stage procedure provides both better estimation accuracy and convergence characteristics.

Determination of the Spacecraft Position by Single Station Antenna Tracking Data

In general, for the orbit determination purpose, the Kalman filtering technique is used. Accuracy of the Kalman filter depends on the measurement accuracy significantly. Therefore, it is important to derive formulas for the accuracy (variances) of indirect x-y-z position measurements.

Spacecraft coordinates are determined with single point (ground station) by the help of this method and difficult calculations are not needed. The following formulas are used to calculate spacecraft's coordinates in the UEN coordinate frame [18, 21];

$$\begin{aligned} \rho_N &= \rho \cos\beta \cos\alpha \\ \rho_E &= \rho \cos\beta \sin\alpha \\ \rho_U &= \rho \sin\beta \end{aligned} \quad (26)$$

Range, azimuth and elevation angles are determined by radiolocation measurements.

After accomplishing the successive coordinate transformations on the measured position between UEN and Earth-Centered-Earth-Fixed (ECEF) and between ECEF and ECI [18, 22], the calculated indirect position measurements x, y, z of the spacecraft in ECI is obtained as a function of the real measurements ρ, α, β in UEN coordinate frame as follows;

$$\begin{aligned} x &= \rho \cos\theta \cos\lambda \sin\beta - \rho \cos\theta \sin\lambda \cos\beta \cos\alpha \\ &\quad - \rho \sin\theta \cos\beta \sin\alpha + R_E \cos\lambda \cos\theta \\ y &= \rho \sin\theta \cos\lambda \sin\beta - \rho \sin\theta \sin\lambda \cos\beta \cos\alpha \\ &\quad + \rho \cos\theta \cos\beta \sin\alpha + R_E \cos\lambda \sin\theta \\ z &= \rho \sin\lambda \sin\beta + \rho \cos\lambda \cos\beta \cos\alpha + R_E \sin\lambda \end{aligned} \quad (27)$$

Variance Analysis of the Position Errors

In this sub-section, formulas for the error analysis of the spacecraft's orbit determination will be presented. Variances of orbit parameters are chosen as the accuracy criteria, and the accuracy of algorithm (27) is considered for construction of the covariance matrix of measurement noise, R in EKF.

The method to calculate the accuracy of spacecraft's position enables determining possibility characteristics of the position data's error depending on the type of ground station, location of ground station, and measurement accuracy of navigation parameters [21, 23]. Establishing ground stations optimally and choosing the most accurate position determining area are possible as a result of the needed calculations. In some cases, due to the difficulty of the calculations and the need for huge preparations, approximate accuracy values of the coordinates are used to determine the accuracy of the position data. The coordinates of the spacecraft are calculated by formulas (27). Because, x, y, z coordinates of the spacecraft are non-linear functions of the navigation parameters, range (ρ), azimuth angle (α), elevation angle (β) measured by single ground station antenna, by accepting that the navigation parameters' measurement errors are independent and errors of parameters θ, λ, R_E are negligible, the variances of coordinates' calculation errors are determined as [21];

$$\sigma_x^2 = \left(\frac{\partial x}{\partial \rho}\right)^2 \sigma_\rho^2 + \left(\frac{\partial x}{\partial \alpha}\right)^2 \sigma_\alpha^2 + \left(\frac{\partial x}{\partial \beta}\right)^2 \sigma_\beta^2$$

$$\sigma_y^2 = \left(\frac{\partial y}{\partial \rho}\right)^2 \sigma_\rho^2 + \left(\frac{\partial y}{\partial \alpha}\right)^2 \sigma_\alpha^2 + \left(\frac{\partial y}{\partial \beta}\right)^2 \sigma_\beta^2 \qquad (28)$$

$$\sigma_z^2 = \left(\frac{\partial z}{\partial \rho}\right)^2 \sigma_\rho^2 + \left(\frac{\partial z}{\partial \alpha}\right)^2 \sigma_\alpha^2 + \left(\frac{\partial z}{\partial \beta}\right)^2 \sigma_\beta^2$$

where σ_x^2, σ_y^2, σ_z^2 are the variances of the calculated x, y, z coordinates' errors, respectively; σ_ρ^2, σ_α^2, σ_β^2 are the variances of the measurement errors of navigation parameters ρ, α, β, respectively. The expressions for the partial derivatives in (28) are presented in Appendix.

Estimation of Satellite Position and Velocity on Linear Position Measurements

In this approach, the measurement vector can be presented as

$$Z^* = \begin{bmatrix} x & y & z \end{bmatrix}^T \qquad (29)$$

This new approach aims to simplify the measurement matrix as

$$H = \begin{bmatrix} 1 & 0 & 0 & 0 & 0 & 0 \\ 0 & 0 & 1 & 0 & 0 & 0 \\ 0 & 0 & 0 & 0 & 1 & 0 \end{bmatrix} \qquad (30)$$

by deriving the components of the discrete measurement vector Z^*_k from the directly measured range, azimuth, and elevation.

By substituting the expression in Equations (29) and (30) in the EKF in Equations (18)-(25) instead of Z and $\nabla h_X(k+1)$, and by taking into account that $h\ \hat{X}(k+1/k) = H\hat{X}(k+1/k)$, the following EKF for estimation of the spacecraft's position and velocity based on indirect linear x-y-z measurements can be obtained:

The state estimation

$$\hat{X}(k+1/k+1) = \hat{X}(k+1/k) + K(k+1)\Delta(k+1) \qquad (31)$$

the innovation

$$\Delta(k+1) = Z^*(k+1) - H\hat{x}(k+1/k) \qquad (32)$$

the state prediction

$$\hat{X}(k+1/k) = f(\hat{X}(k/k)) \qquad (33)$$

the Kalman gain

$$K(k+1) = P(k+1/k)H^T(k+1)\begin{bmatrix} H(k+1)P(k+1/k)H^T(k+1) \\ +R(k+1) \end{bmatrix}^{-1} \quad (34)$$

the covariance estimation

$$P(k+1/k+1) = [I - K(k+1)H(k+1)]P(k+1/k) \quad (35)$$

the covariance prediction

$$P(k+1/k) = F_X P(k/k) F_X^T + GQ(k)G^T \quad (36)$$

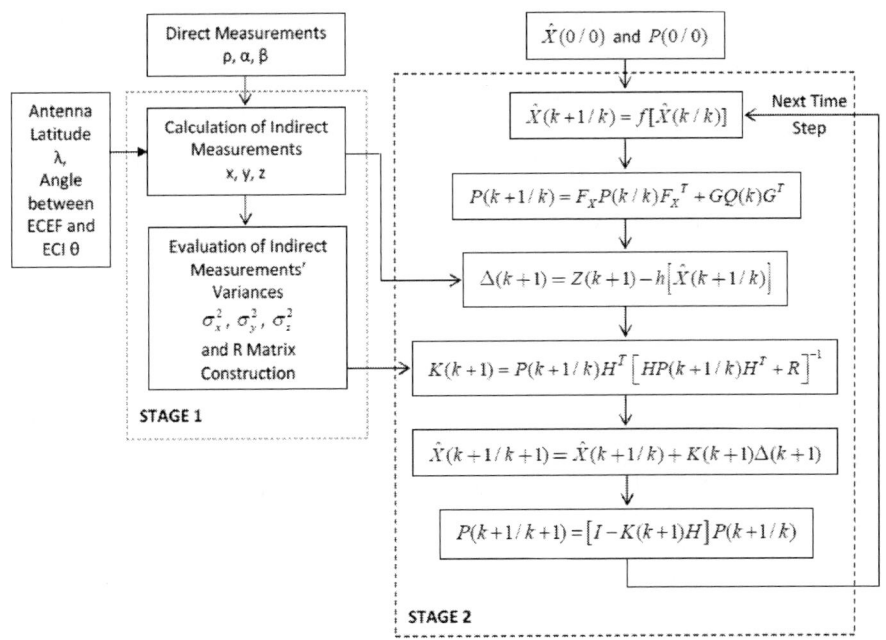

Figure 1. The structural scheme of the proposed EKF.

The expression for F_x is the same as in the EKF in Equations (18)-(25).

The structural scheme of the proposed EKF for estimation of spacecraft's position and velocity on indirect linear x-y-z measurements is given in Figure 1.

SIMULATIONS

Estimation by Using Range-Azimuth-Elevation Measurements Directly

This subsection will give the steps to be followed to simulate the estimation of the position and velocity of a geostationary satellite by using its range, azimuth angle, and elevation angle with respect to a ground antenna.

The ground antenna is assumed to be located in the north-western part of Turkey, so its latitude angle is taken as

$$\lambda = 40°.$$

The standard deviation values of the range, azimuth, and elevation measurements are accepted to be

$$\sigma_\rho = 10 \text{ m}$$
$$\sigma_\alpha = 0.01°,$$
$$\sigma_\beta = 0.01°$$

respectively. Because the measurement vector of the extended Kalman filter consists of the measured range, azimuth, and elevation values

$$Z = \begin{bmatrix} \rho & \alpha & \beta \end{bmatrix}^T, \tag{37}$$

the measurement covariance matrix can be written as follows

$$R = \begin{bmatrix} \sigma_\rho^2 & 0 & 0 \\ 0 & \sigma_\alpha^2 & 0 \\ 0 & 0 & \sigma_\beta^2 \end{bmatrix}. \qquad (38)$$

The elements of Z are measured in the Up-East-North (UEN) coordinate system. Because the position and velocity components of the spacecraft, which are the states of the estimation problem, are defined in the Earth-Centered Inertial (ECI) coordinate system, the measured variables has to be expressed as function of the states.

The range vector's components can be written in ECI coordinate frame as follows;

$$\begin{bmatrix} \rho_x \\ \rho_y \\ \rho_z \end{bmatrix} = \begin{bmatrix} x - R_E \cos \lambda \cos \theta \\ y - R_E \cos \lambda \sin \theta \\ z - R_E \sin \lambda \end{bmatrix}. \qquad (39)$$

Here; x, y, z are the inertial position components of the spacecraft; R_E is the Earth's radius at equator; and θ is the rotation angle of the Earth-Centered Earth-Fixed (ECEF) reference system around the Earth's rotation axis with respect to the ECI coordinate system. It is calculated according to

$$\theta = 280.46061837 + 360.98564736628 \, d_{2000} \qquad (40)$$

in degrees. d_{2000} is the duration from the epoch at the beginning of the year 2000 till the starting moment of the estimation process and calculated by the formula

$$d_{2000} = 367y - \text{INT}\left\{ \frac{7\{y + \text{INT}[(m+9)/12]\}}{4} \right\}$$

$$+ \text{INT}\left\{ \frac{275m}{9} \right\} + \frac{h + \min/60 + s/3600}{24} + d - 730531.5$$

in days [16]. The starting moment for this work is input to the simulation as y=2011, m=12, d=18, h=12, min=0, s=0.

The transformation matrix from the ECI system to the ECEF system is

$$T^{ECI \to ECEF} = \begin{bmatrix} \cos\theta & \sin\theta & 0 \\ -\sin\theta & \cos\theta & 0 \\ 0 & 0 & 1 \end{bmatrix}, \quad (41)$$

whereas the transformation matrix from the ECEF system to the UEN system is as follows;

$$T^{ECEF \to UEN} = \begin{bmatrix} \cos\lambda & 0 & \sin\lambda \\ 0 & 1 & 0 \\ -\sin\lambda & 0 & \cos\lambda \end{bmatrix}. \quad (42)$$

Therefore, the quantities measured in the UEN system can be related to the position components defined in the ECI system through the following transformation;

$$\begin{bmatrix} \rho_U \\ \rho_E \\ \rho_N \end{bmatrix} = \left(T^{ECEF \to UEN}\right)\left(T^{ECI \to ECEF}\right) \begin{bmatrix} \rho_x \\ \rho_y \\ \rho_z \end{bmatrix}. \quad (43)$$

According to Equations (10)-(12), the measurement vector is related to the state vector through a nonlinear measurement vector function, which is defined in the discrete measurement equation of the extended Kalman filter as

$$Z_k = \begin{bmatrix} \rho_k \\ \alpha_k \\ \beta_k \end{bmatrix} = h(X_k) + V_k. \quad (44)$$

Equation (44) is implemented in the simulation as

$$Z_k = \begin{bmatrix} \sqrt{\rho_U^2 + \rho_E^2 + \rho_N^2}\Big|_k \\ \tan^{-1} \dfrac{\rho_E}{\rho_N}\Big|_k \\ \sin^{-1} \dfrac{\rho_U}{\rho}\Big|_k \end{bmatrix} + randn \begin{bmatrix} \sigma_\rho \\ \sigma_\alpha \\ \sigma_\beta \end{bmatrix}. \qquad (45)$$

The measurement matrix H of the extended Kalman filter is derived by calculating the Jacobian of the $h(x_k, y_k, z_k)$, which has the form of

$$H = \dfrac{\partial h(X)}{\partial X}\bigg|_{X=\hat{X}_k} = \begin{bmatrix} \dfrac{\partial \hat{\rho}_k}{\partial \hat{x}} & 0 & \dfrac{\partial \hat{\rho}_k}{\partial \hat{y}} & 0 & \dfrac{\partial \hat{\rho}_k}{\partial \hat{z}} & 0 \\ \dfrac{\partial \hat{\alpha}_k}{\partial \hat{x}} & 0 & \dfrac{\partial \hat{\alpha}_k}{\partial \hat{y}} & 0 & \dfrac{\partial \hat{\alpha}_k}{\partial \hat{z}} & 0 \\ \dfrac{\partial \hat{\beta}_k}{\partial \hat{x}} & 0 & \dfrac{\partial \hat{\beta}_k}{\partial \hat{y}} & 0 & \dfrac{\partial \hat{\beta}_k}{\partial \hat{z}} & 0 \end{bmatrix} \qquad (46)$$

according to the definition of the state vector. The terms in the H matrix are derived in [16] and are presented in Appendix.

Estimation by Using Indirect x-y-z Measurements

The following relation between the inertial position components of the spacecraft and $\begin{bmatrix} \rho_x & \rho_y & \rho_z \end{bmatrix}^T$ is obvious from Equation (39) [24];

$$\begin{bmatrix} x \\ y \\ z \end{bmatrix} = \begin{bmatrix} \rho_x \\ \rho_y \\ \rho_z \end{bmatrix} + R_E \begin{bmatrix} \cos\lambda \cos\theta \\ \cos\lambda \sin\theta \\ \sin\lambda \end{bmatrix}. \qquad (47)$$

By performing the inverse of the transformation given in Equation (43), $\begin{bmatrix} \rho_x & \rho_y & \rho_z \end{bmatrix}^T$ can be expressed in terms of the variables measured in the UEN coordinate system. Because the related transformation matrices defined in Equations (41) and (42) are orthogonal, their inverses are equal to their transposes, so

$$\begin{bmatrix} \rho_x \\ \rho_y \\ \rho_z \end{bmatrix} = \left(T^{ECI \to ECEF} \right)^T \left(T^{ECEF \to UEN} \right)^T \begin{bmatrix} \rho_U \\ \rho_E \\ \rho_N \end{bmatrix} \qquad (48)$$

holds. By substituting Equation (48) into Equation (47), the elements of the measurement vector in Equation (29) that will be input to the extended Kalman filter can be obtained as shown in Equation (27).

Resulting from Equation (28), the measurement covariance matrix becomes

$$R = \begin{bmatrix} \sigma_x^2 & 0 & 0 \\ 0 & \sigma_y^2 & 0 \\ 0 & 0 & \sigma_z^2 \end{bmatrix}. \qquad (49)$$

The initial state vector is input to both filters as

$$X_0^T = \begin{bmatrix} -54513753 \text{ m} \\ 3975.20 \text{ m/s} \\ -54513753 \text{ m} \\ -3975.20 \text{ m/s} \\ 500000 \text{ m} \\ -50 \text{ m/s} \end{bmatrix}$$

The system noise covariance matrix is selected as

$$Q = \begin{bmatrix} 0 & 0 & 0 & 0 & 0 & 0 \\ 0 & 0.1 \times \Delta t & 0 & 0 & 0 & 0 \\ 0 & 0 & 0 & 0 & 0 & 0 \\ 0 & 0 & 0 & 0.1 \times \Delta t & 0 & 0 \\ 0 & 0 & 0 & 0 & 0 & 0 \\ 0 & 0 & 0 & 0 & 0 & 0.1 \times \Delta t \end{bmatrix} \quad (50)$$

with $\Delta t = 1$ second being the simulation step.

Results

After one day long simulation, the three components of the GEO satellite's position is estimated by both filters as shown in Figure 2 in comparison.

As seen from Figure 2 (a, b), the transient responses by the indirect method is more rapid than the ones by the direct method. The method based on indirect measurements provides an unnoticeably short transient duration while the method based on direct measurements makes the estimated states converge to the real states after 0.4 days. The results for the three components of the velocity are compared in Figure 3 (a, b).

Absolute estimation error graphs for position and velocity components are presented in Figures 4 (a, b) and 5 (a, b).

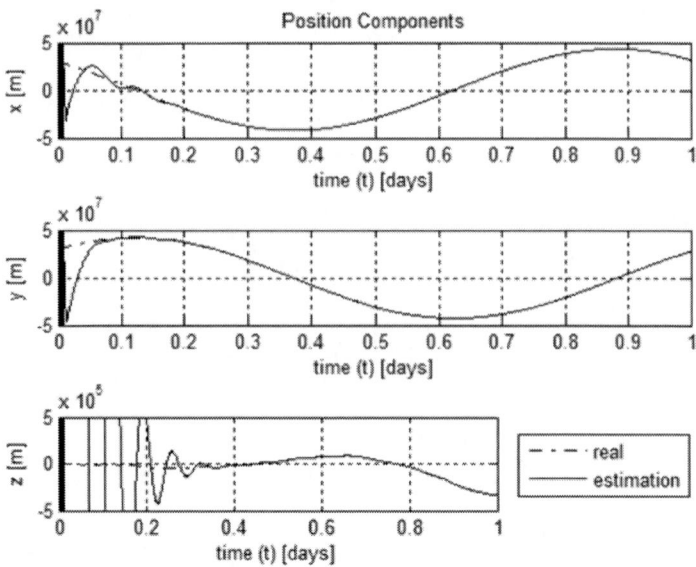

Figure 2a. Estimation of satellite's position components by using direct range-azimuth-elevation.

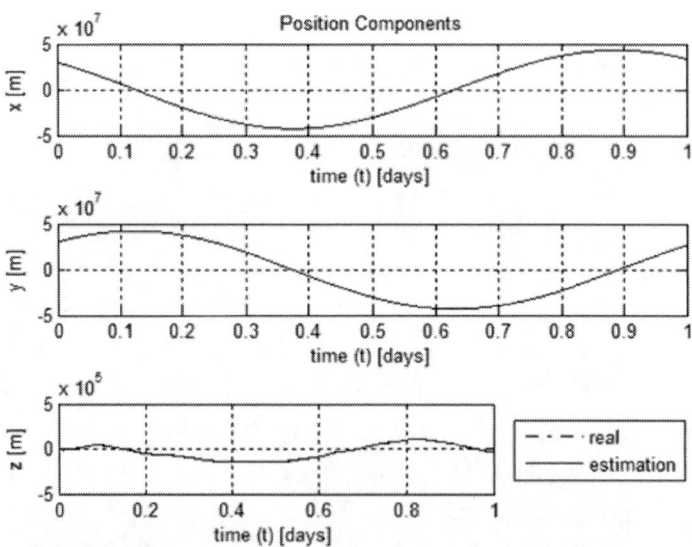

Figure 2b. Estimation of satellite's position components by using indirect x-y-z measurements.

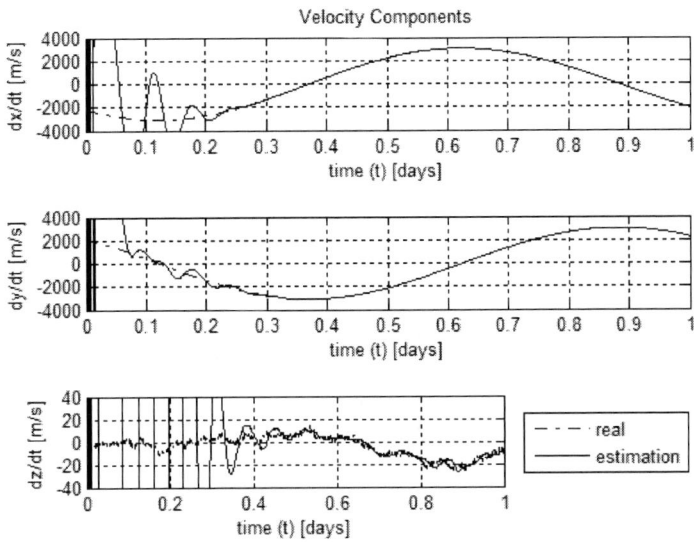

Figure 3a. Estimation of satellite's velocity components by using direct range-azimuth-elevation.

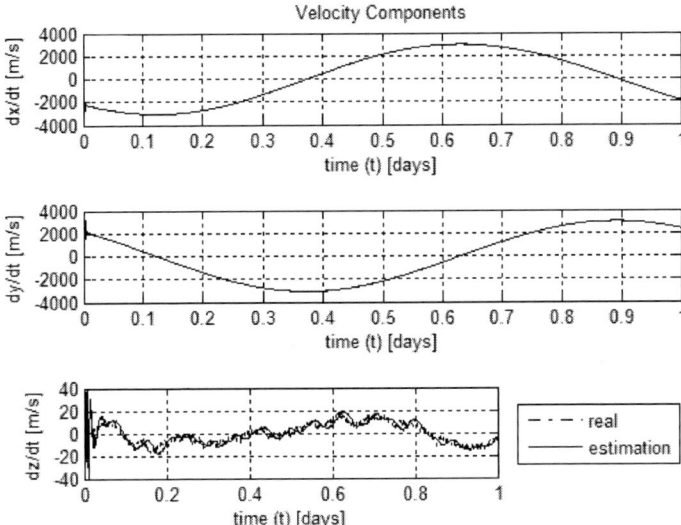

Figure 3b. Estimation of satellite's velocity components by using indirect x-y-z measurements.

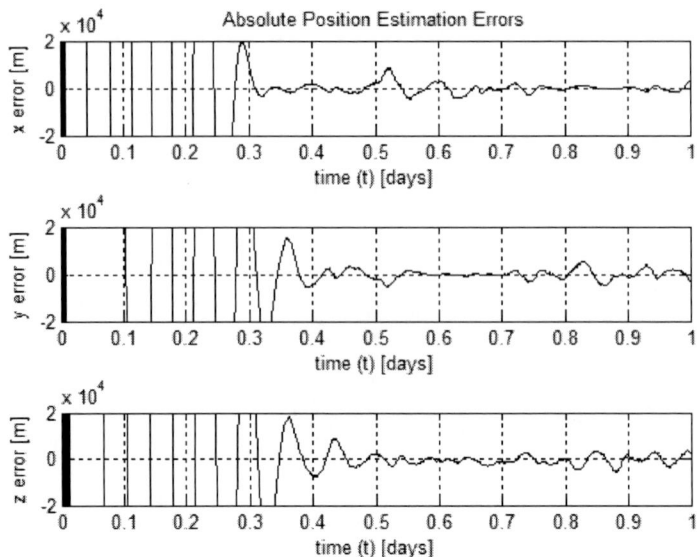

Figure 4a. Absolute estimation errors of satellite's position components by using direct range-azimuth-elevation.

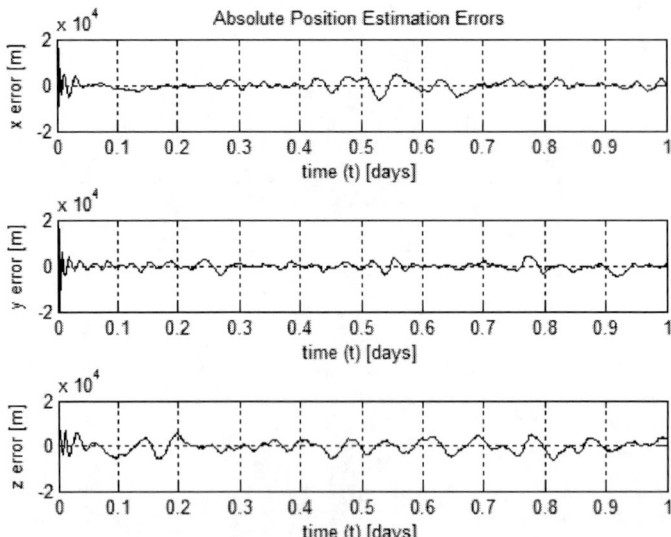

Figure 4b. Absolute estimation errors of satellite's position components by using indirect x-y-z measurements.

EKF-Based Orbit Determination Using Single Ground Tracking ...

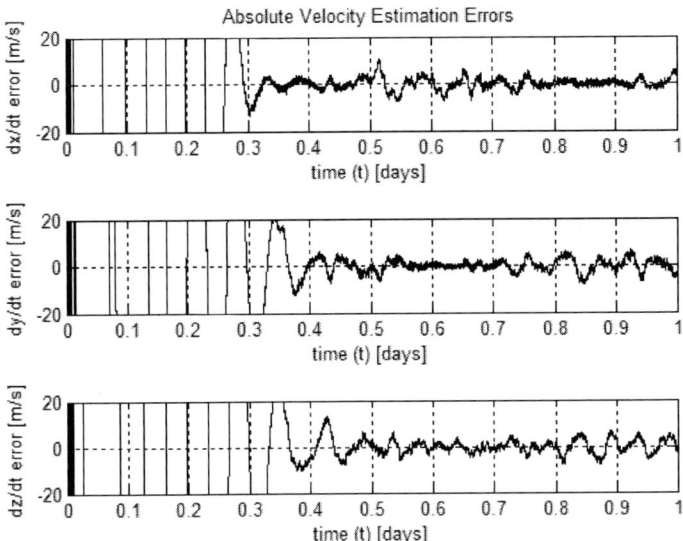

Figure 5a. Absolute estimation errors of satellite's velocity components by using direct range-azimuth-elevation.

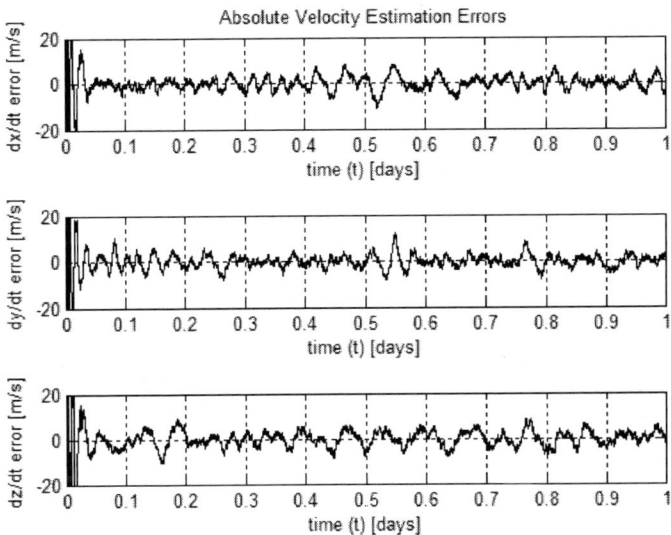

Figure 5b. Absolute estimation errors of satellite's velocity components by using indirect x-y-z measurements.

Table 1, 2, and 3 list absolute estimation errors of the position and velocity components in the x-direction for one day.

Table 1. Absolute position and velocity estimation errors in the x-direction for one day

Time (t) [s]	x [m]		dx/dt [m/s]	
	by *direct* measurements	by *indirect* measurements	by *direct* measurements	by *indirect* measurements
0	8.43283e+07	8.43283e+07	6.14931e+03	6.14931e+03
5760	6.69962e+06	9.65498e+01	2.38057e+03	3.45131e-03
11520	2.54346e+06	2.87117e+03	5.49056e+02	2.05045
17280	4.92154e+05	1.09247e+03	1.96014e+02	1.12225
23040	5.25276e+04	1.72885e+03	3.39979e+01	2.13969
28800	6.93143e+02	2.80410e+02	2.67862	2.21168
34560	1.75610e+03	2.51880e+01	7.38972e-02	1.07012
40320	1.66455e+03	9.56961e+02	1.24399e-01	7.43421
46080	2.91718e+03	6.15691e+03	4.14769	2.00783
51840	2.63329e+03	2.58262e+03	1.97434	3.55080e-01
57600	1.80425e+03	3.98546e+03	3.67889	2.51727e-01
63360	5.64824e+02	1.20116e+03	1.76266	3.13683
69120	8.69758e+02	1.72847e+03	6.90751e-01	4.94055
74880	2.20715e+02	7.58029e+01	7.86441e-01	1.49788
80640	1.06234e+03	1.60644e+03	9.32749e-01	1.05545
86400	3.15542e+03	4.79683e+01	2.83696	4.07462
86401	3.09944e+03	5.82050e+01	2.79168	4.11315
rms	*2.70603e+08*	*2.86905e+05*	*3.20052e+07*	*5.82060e+02*

Table 2. Absolute position and velocity estimation errors in the y-direction for one day

Time (t) [s]	y [m]		dy/dt [m/s]	
	by *direct* measurements	by *indirect* measurements	by *direct* measurements	by *indirect* measurements
0	8.43283e+07	8.43283e+07	6.14931e+03	6.14931e+03
5760	2.50717e+06	9.30298e+02	8.25098e+02	1.43124e-01
11520	3.83060e+05	1.30281e+03	3.97852e+01	5.08396
17280	3.55064e+05	5.28096e+01	8.24753e+01	3.04241
23040	1.01486e+05	3.99194e+03	3.69319e+01	4.16077
28800	2.18221e+04	4.78058e+02	9.67202	2.05557e-01
34560	4.51402e+03	4.38394e+01	1.14254	2.52513
40320	2.61902e+03	8.42455e+02	1.36199	2.38057

Time (t) [s]	y [m] by *direct* measurements	y [m] by *indirect* measurements	dy/dt [m/s] by *direct* measurements	dy/dt [m/s] by *indirect* measurements
46080	1.48281e+03	2.09594e+03	2.07789	7.09309
51840	6.97470e+01	1.53531e+02	1.07287	2.17496e-01
57600	4.89346e+02	1.27893e+03	3.03397	9.49922e-01
63360	7.80449e+02	5.53972e+02	2.22320	2.02459
69120	1.89461e+03	3.27544e+03	2.23434	9.57846e-02
74880	4.49307e+03	3.56413e+02	5.06469e-01	5.62489e-01
80640	3.02441e+03	2.21629e+03	2.71849e-01	3.54653
86400	2.77620e+03	6.48945e+02	3.29458	1.15459
86401	2.71609e+03	6.28170e+02	3.21705	9.42694e-01
rms	*2.45636e+08*	*2.86937e+05*	*2.60875e+07*	*1.10950e+03*

Table 3. Absolute position and velocity estimation errors in the z-direction for one day

Time (t) [s]	z [m] by *direct* measurements	z [m] by *indirect* measurements	dz/dt [m/s] by *direct* measurements	dz/dt [m/s] by *indirect* measurements
0	5.0e+05	5.0e+05	5.0e+10	5.0e+01
5760	1.26784e+06	7.15343e+02	1.47811e+04	3.17939e-02
11520	1.91195e+06	1.30523e+03	1.98631e+03	5.59505
17280	6.08200e+05	5.27576e+03	1.45332e+02	6.58037e-01
23040	1.59818e+05	2.60714e+03	1.71592e+01	2.78270
28800	3.52866e+04	1.25023e+01	6.09447	3.29488
34560	7.00217e+03	2.32208e+03	3.51896	1.63108
40320	3.32960e+03	1.21133e+03	1.80926	5.63550
46080	7.14924e+02	1.56378e+03	4.26569	4.43837
51840	9.27235e+02	3.30606e+03	1.17653	1.51322
57600	2.26908e+03	9.08818e+02	2.97171	2.41935
63360	2.04786e+03	1.42310e+01	5.16302e-01	1.20061e-01
69120	1.15047e+02	1.49579e+03	6.81926e-01	4.76505
74880	1.08691e+03	1.46393e+03	5.67427	5.28569
80640	1.26788e+03	4.03530e+02	4.98374	1.81085
86400	2.68437e+03	2.91230e+03	1.37187	1.24097
86401	2.72581e+03	2.90765e+03	1.30607	1.35941
rms	*2.44538e+08*	*3.09289e+03*	*3.74278e+07*	*5.12159*

The last rows of the tables include calculated root mean squares (rms) of the errors. The faster transient response characteristic by estimation using indirect measurements results in highly lower rms values after one day simulation.

In addition, the normalized average errors (NAEs) obtained by both methods are plotted and presented in Figures 6 and 7 for position

$$NAEP_k = \sqrt{(x_k - \hat{x}_k)^2 + (y_k - \hat{y}_k)^2 + (z_k - \hat{z}_k)^2} \qquad (51)$$

and velocity components, respectively.

$$NAEV_k = \sqrt{(u_k - \hat{u}_k)^2 + (v_k - \hat{v}_k)^2 + (w_k - \hat{w}_k)^2} \qquad (52)$$

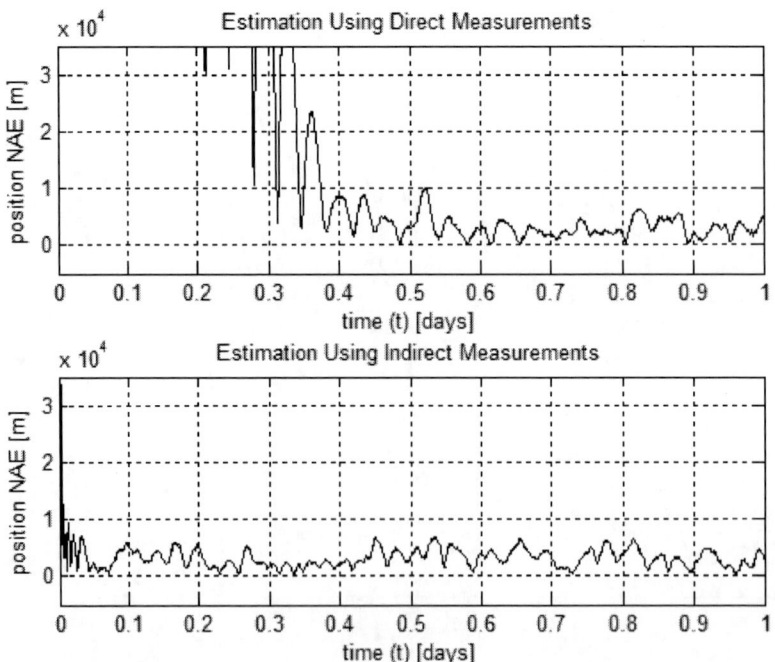

Figure 6. Normalized average position errors in comparison for one day.

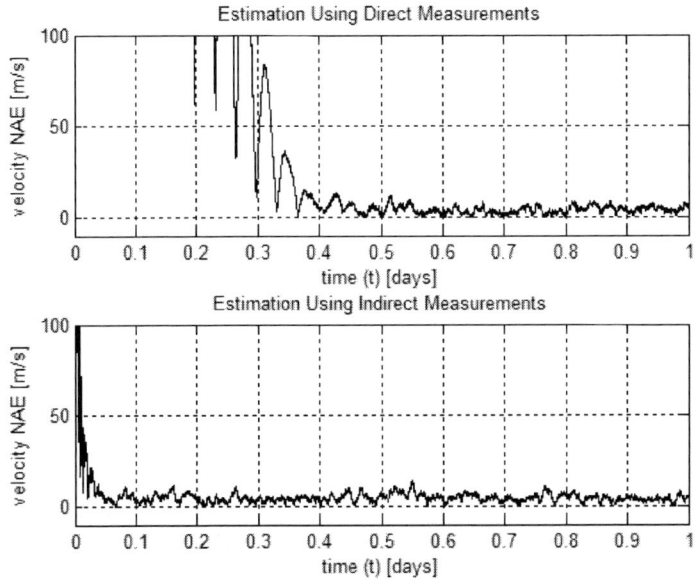

Figure 7. Normalized average velocity errors in comparison for one day.

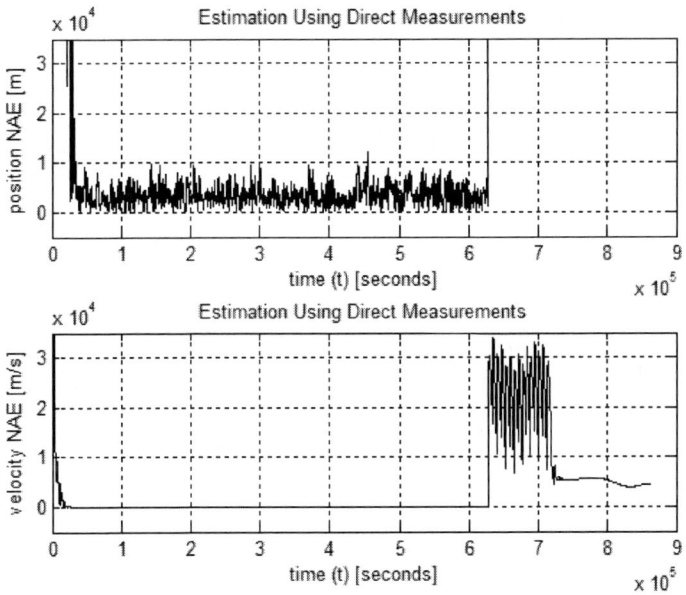

Figure 8. Normalized average position and velocity errors in comparison for ten days.

As seen from Figures 8 and 9, singularities are observed periodically for the direct method during ten day simulation.

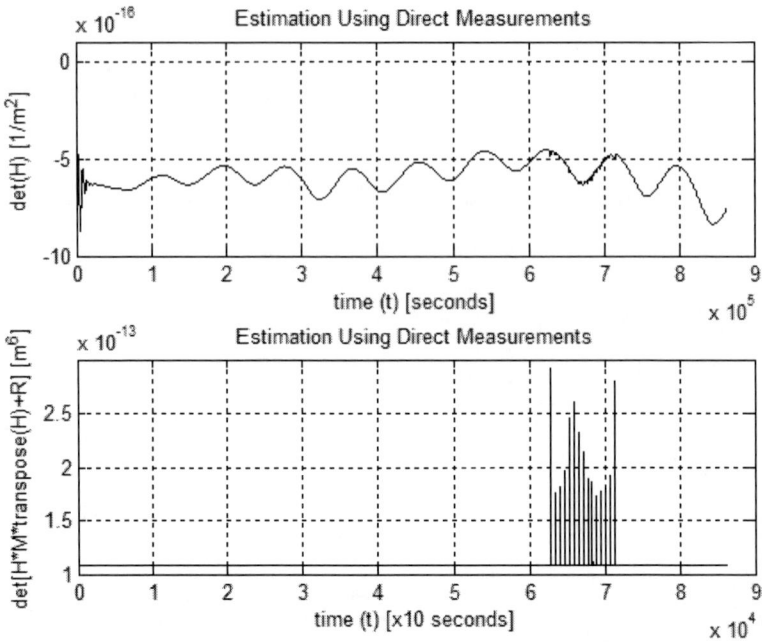

Figure 9. Determinants of the measurement and innovation covariance matrices.

The second method does not lead to singularities, which is another superiority of the indirect method [25]. According to related investigations, the determinant of H is very low, and it takes minimum values around singularity points. The Jacobian H matrix is well known to cause singularity. In the proposed method, the measurement matrix is not a Jacobian, but a constant matrix, which is an important advantage of the designed method.

If the simulation duration is increased to ten days, the root mean squares (rms's) during the steady state can also be calculated. Table 4 tabulates absolute estimation errors of the position and velocity components in the x-direction along ten days. Its last row includes the rms's obtained for the tenth day whereas the previous row belongs to rms values for the whole interval.

Table 4. Absolute position and velocity estimation errors in the x-direction for ten days

Time (t) [s]	x [m]		dx/dt [m/s]	
	by *direct* measurements	by *indirect* measurements	by *direct* measurements	by *indirect* measurements
0	8.43283e+07	8.43283e+07	6.14931e+03	6.14931e+03
57600	2.31993e+03	1.20161e+03	5.30573e-01	2.27577
115200	3.98449e+02	1.15710e+03	7.83430e-01	1.78448
172800	8.86766e+03	1.67772e+03	8.26823e-01	5.37356
230400	4.86164e+02	2.06963e+03	2.90473	2.50807
288000	5.80797e+02	2.36978e+02	7.27929e-01	2.43195e-01
345600	5.50642e+02	2.54444e+03	4.94946e-01	1.32380
403200	8.84734e+02	2.96573e+03	2.55319	8.51663e-01
460800	2.37410e+02	3.68062e+03	5.11173e-01	6.96170e-01
518400	7.26187e+03	2.55135e+03	1.36965	4.81822
576000	8.85256e+02	7.29798e+02	5.59580e-01	3.15474e-01
633600	4.30350e+02	1.48246e+03	1.35474	1.32336
691200	9.67203	9.96298e+02	2.88067	1.17490
748800	3.94113e+02	6.14029e+02	6.97798e-01	2.00031
806400	4.55408e+03	1.92381e+03	1.13414	3.47123
864000	2.40605e+03	1.96325e+02	1.20348e+01	4.00370e-01
rms	*3.47422e+09*	*9.09933e+04*	*2.74313e+08*	*9.44564e+02*
rms_9-10	*9.87660e+02*	*7.71864e+02*	*1.10174*	*1.01151*

The estimation accuracy of the indirect method is better, which can be clearly observed if the estimation duration is long enough as seen from the last row of Table 4. The tables for the position and velocity components in y- and z- direction are given in Tables 5 and 6, respectively.

It is also seen that the direct method is sensitive to initialization parameters' values therefore if the appropriate parameters are not selected, the outputs of the filter may diverge. The indirect method is able to make the filter output converge.

Lastly, the variation of the diagonal elements of the covariance matrix P corresponding to position and velocity components can be seen in Figures 10 and 11, respectively.

Table 5. Absolute position and velocity estimation errors in the y-direction for ten days

Time (t) [s]	y [m]		dy/dt [m/s]	
	by *direct* measurements	by *indirect* measurements	by *direct* measurements	by *indirect* measurements
0	8.43283e+07	8.43283e+07	6.14931e+03	6.14931e+03
57600	1.04274e+03	8.49235e+02	5.69674e-01	5.39672e-01
115200	5.70241e+03	5.12023e+03	8.36677e-01	4.54166e-01
172800	5.59529e+03	1.98917e+03	6.29693e-01	1.83333
230400	5.11218e+02	3.99471e+02	3.49009	5.74927
288000	5.07445e+03	2.80063e+03	6.70745	3.50381
345600	5.78284e+02	8.71338e+02	6.68812e-01	1.85272
403200	1.47241e+03	1.56552e+02	2.78933	1.12334
460800	5.17312e+02	1.31656e+03	3.05214	6.61645
518400	1.54100e+03	8.06086e+02	9.60842e-01	1.99171
576000	1.22081e+03	1.14061e+03	1.98690	6.54478e-01
633600	1.90507e+02	1.93856e+02	3.69955	1.54686
691200	5.80728e+02	1.27123e+03	1.91107	3.05397
748800	3.53143e+03	2.13096e+03	4.23488	5.23069
806400	7.31242e+03	1.13850e+03	3.16105e-02	3.31165e-01
864000	1.94555e+02	4.76463e+03	3.15324	1.40975
rms	*2.69338e+09*	*9.17328e+04*	*2.87800e+08*	*1.72226e+03*
rms_9-10	*9.48701e+02*	*9.42002e+02*	*1.16375*	*1.19702*

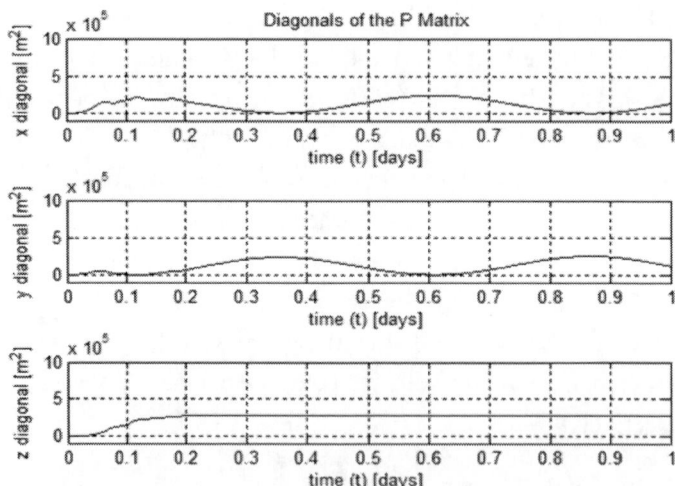

Figure 10a. Diagonal elements of P corresponding to position components obtained by using direct range-azimuth-elevation.

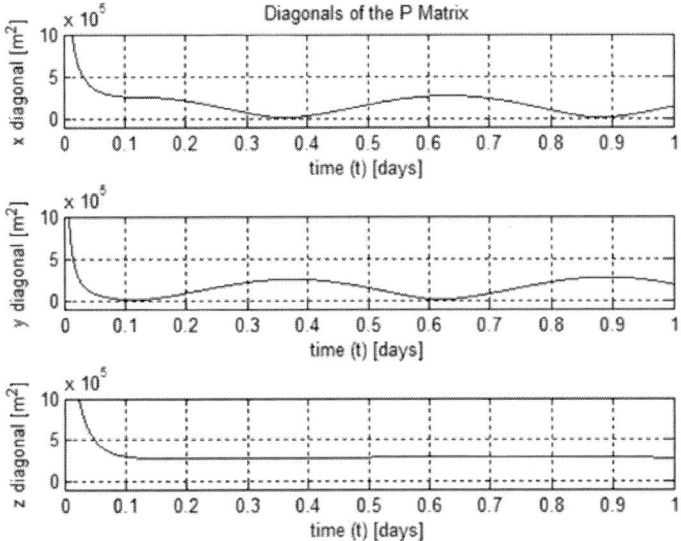

Figure 10b. Diagonal elements of P corresponding to position components obtained by using indirect x-y-z measurements.

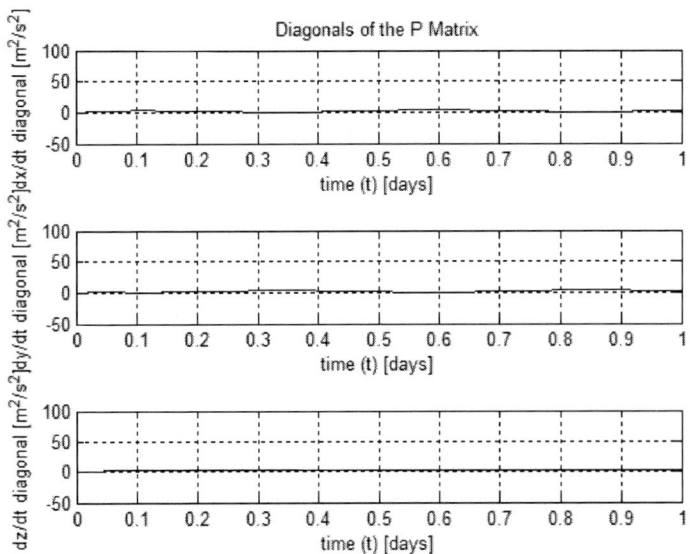

Figure 11a. Diagonal elements of P corresponding to velocity components obtained by using direct range-azimuth-elevation.

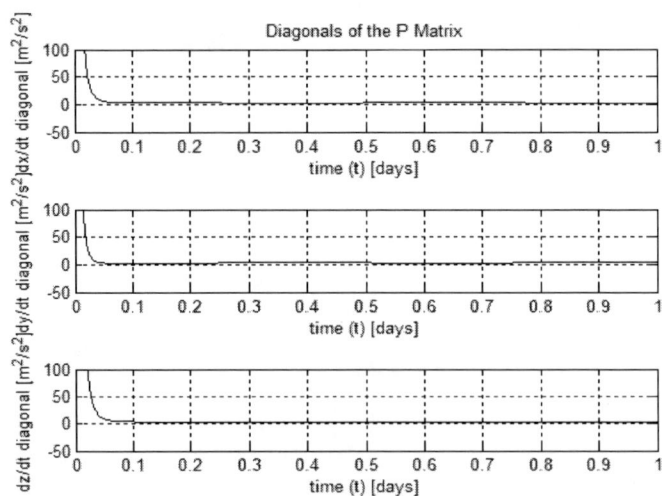

Figure 11b. Diagonal elements of P corresponding to velocity components obtained by using indirect x-y-z measurements.

Table 6. Absolute position and velocity estimation errors in the z-direction for ten days

Time (t) [s]	z [m]		dz/dt [m/s]	
	by *direct* measurements	by *indirect* measurements	by *direct* measurements	by *indirect* measurements
0	5.0e+05	5.0e+05	5.0e+01	5.0e+01
57600	2.87679e+02	5.99027e+03	4.42365	3.98818
115200	1.97563e+03	2.87126e+03	1.36521e-01	2.84298
172800	1.81382e+03	4.53682e+03	1.04361	1.39933
230400	1.27746e+03	1.25766e+04	7.84436e-01	1.58352
288000	1.48937e+03	5.20394e+02	3.86585	5.56317e-01
345600	3.65407e+03	1.86332e+03	1.14131	1.60568
403200	4.62094e+03	1.87526e+02	5.16007	8.26221
460800	4.54722e+03	5.66024e+03	1.40102	6.39945
518400	2.18436e+03	1.22153e+03	3.44734	2.12925e-01
576000	8.88493e+02	1.41095e+03	7.61360	2.90691
633600	4.72758e+03	1.98422e+03	7.87972e-01	6.31104e-01
691200	5.38604e+03	6.39015e+03	3.23679	5.21830
748800	6.74144e+03	5.40706e+01	3.16081	2.59145
806400	7.85012e+03	2.53161e+03	3.40615	1.27010
864000	7.41732e+03	3.58787e+03	3.16769	1.27884
rms	3.00906e+09	3.99689e+03	2.92422e+08	4.34747
rms_9-10	1.53389e+03	1.26959e+03	1.47931	1.28757

It is observed from the last two figures that the variances of the theoretical estimation errors converge to zero more rapidly by the direct method, which is expected because the R matrix is varying in the indirect method whereas it is constant in the direct method. It is a disadvantage of the proposed method.

CONCLUSION

A new extended Kalman filter that uses indirect position measurements to estimate the position and velocity components of spacecraft is developed. Inputting the position components of the satellite in ECI reference frame calculated from real range measurements in UEN reference frame to the filter leads to a linear measurement function. The conventional approach of inputting directly the real range-azimuth-elevation measurements results in a highly nonlinear measurement function. The corresponding Jacobian H matrix consists of complex terms, which increases the computational load significantly. For most of the applications, generation of Jacobians is hard, time consuming, and prone to human errors. Nonetheless, linearization brings about an unstable filter performance when the time step intervals for update are not sufficiently small, which results in filter divergence. Per contra, small time step intervals increase the computational burden due to larger number of Jacobian calculations.

In this study, two-stage estimation of spacecraft's position and velocity by single station antenna tracking data, where the measurement nonlinearities are eliminated, is proposed. In the first stage, the direct nonlinear antenna measurements are transformed to the linear x-y-z coordinate measurements of spacecraft's position, and their errors' variances are evaluated. In the second stage, the solutions of the first stage are improved by EKF for estimation of the spacecraft position and velocity on indirect linear x-y-z measurements.

The results of examined filters are compared through geostationary satellite orbital motion simulations. The method based on indirect measurements provides an unnoticeably short transient duration. The complex content of the measurement matrix in the conventional method causes periodic singularities in simulation results. In addition, the convergence of the filter using direct measurements is highly dependent on the initialization parameters' values due to the nonlinear partial derivatives in the Jacobian measurement matrix. It may diverge for various values of the mentioned parameters whereas the proposed filter is much more stable and makes the convergence possible for a wider range of values. The comparison of the accuracy of both methods shows that the estimation by using indirect measurements reduces the absolute estimation errors. The simulation results show that the proposed two-stage procedure performs both with better estimation accuracy and better convergence characteristics. On the other hand, the variances of the theoretical estimation errors converge to zero more rapidly by the direct method, which is expected because the R matrix is varying in the indirect method whereas it is constant in the direct method. It is a disadvantage of the proposed method.

APPENDICES

Extended Kalman Filter

Consider the class of nonlinear systems driven by white noise with white noise-corrupted mesasurements defined by

$$x(k+1) = \varphi[x(k),k] + w(k)$$
$$z(k) = h[x(k),k] + v(k) \qquad \text{(A-1)}$$

where $x(k+1)$ is the state vector, $z(k)$ is the measurement at time k, $w(k)$ is the system noise, $v(k)$ is the measurement noise, $\phi[x(k),k]$ is the nonlinear state transition function mapping the previous state to the current state, $h[x(k),k]$ is a nonlinear measurement model mapping current state to measurements.

It is assumed that both noise vectors $v(k)$ and $w(k)$ are linearly additive Gaussian, temporally uncorrelated with zero mean, which means

$$E[w(k)] = E[v(k)] = 0, \ \forall k, \qquad \text{(A-2)}$$

with the corresponding covariances

$$E\left[w(i)w^T(j)\right] = Q(i)\delta(ij), \quad E\left[v(i)v^T(j)\right] = R(i)\delta(ij) \qquad \text{(A-3)}$$

where $\delta(ij)$ is the Kronecker symbol.

It is assumed that process and mesasurement noises are uncorrelated, i.e.,

$$E\left[w(i)v^T(j)\right] = 0, \quad \forall i, j. \qquad \text{(A-4)}$$

We will consider a real-time linear Taylor approximation of the system function at the previous state estimate and that of the observation function at the corresponding predicted position. The Kalman Filter obtained will be called the Extended Kalman Filter (EKF).

In this case, filter algorithms are as following [26]

$$\hat{x}(k+1) = \hat{x}(k+1/k) + K(k+1)\{z(k+1) - h[\hat{x}(k+1/k), k+1]\} \qquad \text{(A-5)}$$

One-stage prediction algorithm can be written as

$$\hat{x}(k+1/k) = \varphi[\hat{x}(k), k] \qquad \text{(A-6)}$$

Filter-gain algorithm

$$K(k+1) = P(k+1/k) \frac{\partial h^T[\hat{x}(k+1/k), k+1]}{\partial \hat{x}(k+1/k)}$$
$$\times \left[\begin{array}{l} \frac{\partial h\ [\hat{x}(k+1/k), k+1]}{\partial \hat{x}(k+1/k)} P(k+1/k) \frac{\partial h^T[\hat{x}(k+1/k), k+1]}{\partial \hat{x}(k+1/k)} \\ +R(k) \end{array} \right]^{-1} \qquad \text{(A-7)}$$

Extrapolation error algorithm

$$P(k+1/k) = \frac{\partial \varphi[\hat{x}(k), k]}{\partial \hat{x}(k)} P(k/k) \frac{\partial \varphi^T[\hat{x}(k), k]}{\partial \hat{x}(k)} + Q(k) \qquad \text{(A-8)}$$

Estimation error algorithm

$$P(k+1/k+1) = \left[I - K(k+1)\frac{\partial h\;[\hat{x}(k+1/k), k+1]}{\partial \hat{x}(k+1/k)}\right] P(k+1/k) \qquad \text{(A-9)}$$

The filter expressed by the formulas (A-5)-(A-9) is called the Extended Kalman Filter.

Partial Derivatives Required to Calculate x-y-z Position Errors' Variances

$$\frac{\partial x}{\partial \rho} = \cos\theta \cos\lambda \sin\beta - \cos\theta \sin\lambda \cos\beta \cos\alpha$$
$$- \sin\theta \cos\beta \sin\alpha$$

$$\frac{\partial x}{\partial \alpha} = \cos\theta \sin\lambda \cos\beta \sin\alpha\rho - \sin\theta \cos\beta \cos\alpha\rho$$

$$\frac{\partial x}{\partial \beta} = \cos\theta \cos\lambda \cos\beta\rho + \cos\theta \sin\lambda \sin\beta \cos\alpha\rho$$
$$+ \sin\theta \sin\beta \sin\alpha\rho$$

$$\frac{\partial y}{\partial \rho} = \sin\theta \cos\lambda \sin\beta - \sin\theta \sin\lambda \cos\beta \cos\alpha$$
$$+ \cos\theta \cos\beta \sin\alpha$$

$$\frac{\partial y}{\partial \alpha} = \sin\theta \sin\lambda \cos\beta \sin\alpha\rho + \cos\theta \cos\beta \cos\alpha\rho$$

$$\frac{\partial y}{\partial \beta} = \sin\theta \cos\lambda \cos\beta\rho + \sin\theta \sin\lambda \sin\beta \cos\alpha\rho$$
$$- \cos\theta \sin\beta \sin\alpha\rho$$

$$\frac{\partial z}{\partial \rho} = \sin\lambda \sin\beta + \cos\lambda \cos\beta \cos\alpha$$

$$\frac{\partial z}{\partial \alpha} = -\cos\lambda \cos\beta \sin\alpha\rho$$

$$\frac{\partial z}{\partial \beta} = \sin\lambda \cos\beta\rho - \cos\lambda \sin\beta \cos\alpha\rho$$

Partial Derivatives Required to Calculate Elements of Jacobian Matrix H

$$\frac{\partial \rho}{\partial x} = \left(\rho_U \cos \lambda \cos \theta - \rho_E \sin \theta - \rho_N \sin \lambda \cos \theta \right) / \rho$$

$$\frac{\partial \rho}{\partial y} = \left(\rho_U \cos \lambda \sin \theta + \rho_E \cos \theta - \rho_N \sin \lambda \sin \theta \right) / \rho$$

$$\frac{\partial \rho}{\partial z} = \left(\rho_U \sin \lambda + \rho_N \cos \lambda \right) / \rho$$

$$\frac{\partial \alpha}{\partial x} = \frac{1}{\left(\rho_N^2 + \rho_E^2 \right)} \left(\rho_E \sin \lambda \cos \theta - \rho_N \sin \theta \right)$$

$$\frac{\partial \alpha}{\partial y} = \frac{1}{\left(\rho_N^2 + \rho_E^2 \right)} \left(\rho_E \sin \lambda \sin \theta + \rho_N \cos \theta \right)$$

$$\frac{\partial \alpha}{\partial z} = -\frac{1}{\left(\rho_N^2 + \rho_E^2 \right)} \rho_E \cos \lambda$$

$$\frac{\partial \beta}{\partial x} = \frac{1}{\rho \sqrt{\rho^2 - \rho_U^2}} \left(\rho \cos \lambda \cos \theta - \rho_U \frac{\partial \rho}{\partial x} \right)$$

$$\frac{\partial \beta}{\partial y} = \frac{1}{\rho \sqrt{\rho^2 - \rho_U^2}} \left(\rho \cos \lambda \sin \theta - \rho_U \frac{\partial \rho}{\partial y} \right)$$

$$\frac{\partial \beta}{\partial z} = \frac{1}{\rho \sqrt{\rho^2 - \rho_U^2}} \left(\rho \sin \lambda - \rho_U \frac{\partial \rho}{\partial z} \right)$$

REFERENCES

[1] Pardal, P. C. P. M., Kuga, H. K. and de Moraes, R. V. (2009). A discussion related to orbit determination using nonlinear sigma point Kalman filter. *Mathematical problems in engineering*, 2009 (Article ID 140963): 12 pages. doi:10.1155/2009/140963.

[2] Curti, F. and Longo, F. (2002). Iterative filtering of antenna pointing angles for orbit determination (AAS 01-336). *Advances in the astronautical sciences*, 109 (I): 519-528.

[3] Haedong, K., Hae-Yeon, K., Hwang, Y., Jaehoon, K. and Sun-Byoung, L. (2006). Orbit determination accuracy improvement for geostationary satellite with single station antenna tracking data. *AIAA/AAS Astrodynamics Specialist Conference and Exhibit*, Keystone, Colorado, USA.

[4] Haedong, K., Hae-Yeon, K., Hwang, Y., Jaehoon, K. and Sun-Byoung, L. (2006). Communication, ocean, and meteorological satellite orbit determination analysis considering maneuver scheme. *AIAA/AAS Astrodynamics Specialist Conference and Exhibit*, Keystone, Colorado, USA.

[5] Upadhyay, T., Cotterill, S. and Deaton, A. W. (1993). Autonomous GPS/INS navigation experiment for space transfer vehicle. *IEEE transactions on aerospace and electronic systems*, 29 (3): 772-785. doi:10.1109/7.220929.

[6] Ceccarelli, N., Garulli, A., Giannitrapani, A. and Scortecci, F. (2007). Spacecraft localization via angle measurements for autonomous navigation in deep space. *Proc. of the 17th IFAC Symposium on Automatic Control in Aerospace*, Toulouse, France.

[7] Giannitrapani, A., Ceccarelli, N., Scortecci, F. and Garulli, A. (2011). Comparison of EKF and UKF for spacecraft localization via angle measurements. *IEEE transactions on aerospace and electronic systems*, 47 (1): 75-84. doi:10.1109/TAES.2011.5705660.

[8] Matusewicz, J., Subbarao, K. and Frisbee, J. (2008). Uncertainty characterization of orbital debris using the extended Kalman filter (AAS 07-389). *Advances in the astronautical sciences*, 129 (III): 2163-2188.

[9] Park, E. S., Park, S. Y., Roh, K. M. and Choi, K. H. (2010). Satellite orbit determination using a batch filter based on the unscented transformation. *Aerospace science and technology*, 14 (6): 387-396. doi:10.1016/j.ast.2010.03.007.

[10] Psiaki, M. L. (2017). Gaussian-mixture Kalman filter for orbit determination using angles-only data (Engineering Note). *Journal of guidance, control, and dynamics*, 40 (9): 2339-2345. doi:10.2514/1.G002812.

[11] Jiang, Y., Baoyin, H. and Pengbin, M. (2019). Augmented unbiased minimum-variance input and state estimation for tracking a maneuvering satellite. *Acta astronautica*, 163: 96-107. doi:10.1016/j.actaastro.2018.11.015.

[12] Bruno, O. S. T, Santillo, M. A., Erwin, R. S. and Bernstein, D. S. (2008). Spacecraft tracking using sapmled-data Kalman filters: An illustrative application of extended and unscented estimators. *IEEE control systems magazine*, August: 78-94. doi:10.1109/MCS.2008.923231.

[13] Abdelrahman, M. and Park, S.-Y. (2011). Simultaneous spacecraft attitude and orbit estimation using magnetic field vector measurements. *Aerospace science and technology*, 15 (8): 653-669. doi:10.1016/j.ast.2011.01.006.

[14] Hajiyev, C. and Sofyalı. A. (2013). Two-stage estimation of spacecraft position and velocity via single station antenna tracking data. *Proc. of the 14th International Carpathian Control Conference (ICCC)*, Rytro, Poland, May, 2013, pp.105-110.

[15] Hajiyev, C. and Ata, M. (2015). Unscented Kalman filter based two-stage estimation of spacecraft position and velocity with single station antenna tracking data. In *Proc. 7th International Conference on Recent Advances in Space Technologies*, Istanbul, Turkey, June, 2015, pp. 777-782.

[16] Vallado, D. A. (2007). *Fundamentals of Astrodynamics and Applications*, Third Edition. Published jointly by Microcosm Press, Hawthorne, CA, and Springer, New York, USA.

[17] Hajiyev, C. and Berberoglu, M. I. (2005). EKF based user's position estimation using GNSS measurements. *Proc. of the 3rd Ankara International Aerospace Conference*, Ankara, Turkey.

[18] Crassidis, J. L. and Junkins, J. L. (2004). Optimal Estimation of Dynamic Systems. *Chapman & Hall/CRC Applied Mathematics and Nonlinear Science Series*, CRC Press LLC, USA.

[19] Julier, S. J. and Uhlmann, J. K. (2004). Unscented filtering and nonlinear estimation. *Proc. of the IEEE*, 92 (3): 401-422. doi:10.1109/JPROC.2003.823141.

[20] Julier, S. J., Uhlmann, J. K. and Durrant-Whyte, H. F. (1995). A new approach for filtering nonlinear systems. *Proc. of the American Control Conference*, 3: 1628-1632.

[21] Hajiyev, C. (1999). *Radio Navigation* (in Turkish). Istanbul Technical University Press, Istanbul, Turkey.

[22] Curtis, H. D. (2005). *Orbital Mechanics for Engineering Students*. Elsevier Butterworth-Heinemann.

[23] Hajiyev, C. and Ata, M. (2011). Error analysis of orbit determination for the geostationary satellite with single station antenna tracking data. *Positioning*, 2: 135-144. doi:10.4236/pos.2011.24013.

[24] Sofyali, A. and Hajiyev, C. (2015). Single station antenna–based spacecraft orbit determination via robust EKF against the effect of measurement matrix singularity. *J. Aerosp. Eng.*, 28(1): 04014044-1-04014044-9. doi:10.1061/(ASCE)AS.1943-5525.0000373.

[25] Hajiyev, C. and Sofyali, A. (2018). Spacecraft localization by indirect linear measurements from a single antenna. *Aircraft Engineering and Aerospace Technology*, 90(5): 734-742. doi: 10.1108/AEAT-12-2015-0245.

[26] Sage, A. P. and Mells, J. L. (1971). *Estimation Theory with Applications in Communication and Control*. McGraw-Hill, New York, USA.

In: An Introduction to the Extended ...
Editor: M. Holland

ISBN: 978-1-53618-875-2
© 2020 Nova Science Publishers, Inc.

Chapter 2

KALMAN FILTERS FOR DESCRIPTOR SYSTEMS APPLIED IN DOMAIN OF FAULT TOLERANT CONTROL: A REVIEW

Tigmanshu Patel, M. S. Rao, Jalesh L. Purohit* and V. A. Shah
Dharmsinh Desai University, Nadiad, India

Abstract

Safety, reliability, and dependability of complex industrial processes are of significant importance. Mathematical model based process monitoring and fault tolerant control (FTC) can be employed to address issues related to safe and reliable process operation. The fault diagnosis(FD) and FTC of systems described by ordinary differential equations (ODEs) are relatively well addressed in the literature(Venkatasubramanian et al. 2003b; Venkatasubramanian, Rengaswamy, and Kavuri 2003; Venkatasubramanian et al. 2003b; Isermann 2005; Isermann 1984; Gertler 1991; Miljković 2011; Gao, Ding, and Cecati 2015). However, a large class of systems can be conveniently expressed in descriptor/Differential Algebraic Equation(DAE) form. Descriptor systems tend to possess peculiar properties which in turn demand radically different approaches to address its issues as compared to ODEs. Consequently, the problem of FD and FTC of descriptor systems has received critical attention and continues

---

*Corresponding Author's Email: jalesh.purohit@gmail.com.

to be investigated. In spite, a systematic review on FD and FTC of descriptor systems is by and large missing in literature. Since Kalman filter and its variants are widely used for estimation in diverse domains, we present a review of fault detection, diagnosis and fault tolerant control of descriptor/DAE systems specifically focused on Kalman filter and its variants. Firstly, the domain of FD and FTC is summarized and alternate terminologies are explicitly put forth. Further, the properties and issues of descriptor system are outlined. The approaches for fault detection of descriptor systems are evaluated wherein different Kalman filters are clarified. Sparse work of fault isolation, fault estimation and fault tolerant control of descriptor systems is also outlined. Finally, various examples of descriptor systems across literature are summarized and put forth. The tools and solvers available for descriptor systems are highlighted.

1. INTRODUCTION

Process monitoring and taking appropriate measures to prevent hazards in a timely manner is of paramount importance in order to enhance dependability of industrial processes. In today's world, the control of industrial processes and systems has become increasingly complex due to optimal operation, process integration and inevitable safety of operations. Such modern systems consist of many sub-systems which function in co-ordination with each other to achieve desired performance. These technologically complex systems are sporadically subject to faults which might result into catastrophic consequences(Li 2016) for personnel, system and environment. A fault may occur in a system due to improper operation, usage, mounting, non-adherence to limits, ageing and unsuitable operating environment of system component/s or its subsystems. Consequently, the conventional closed loop control systems are incapable of providing desired performance and stability; under situations where a part of the closed loop system may be subjected to a fault/failure. Additionally, a closed-loop control action may also hide a fault from being observed(Blanke, Staroswiecki, and Wu 2001). The faults within the closed loop system might be caused by sensors, actuators or some structural changes within the system. Such issues apparently make up to 60% of industrial controller problems(Noura et al. 2009). On account of this, the stability and performance of the closed loop system may degrade or may even result in unforeseen consequences; causing economical loss or industrial accidents where damage, injury or loss of life may occur. Hence highly reliable control systems are needed that control these processes in the presence of various faulty scenarios. Such reliability demands fault diagnosis and control which enables a system to take necessary actions in presence of fault/s. The strategy used to control the plant/system in presence of faults depends upon the magnitude, location & type of faults. To ensure that this survey is comprehensive for the reader, basic terminology as well as classification pertaining to FD and FTC is put forth in the following subsection.

1.1. Basic Terminology

1.1.1. Malfunction

A malfunction in a system component is an erratic condition that prevails for a finite amount of time before returning to its nominal operation.

1.1.2. Failure

In contrast, a failure is an irrecoverable condition occurring within a system; that results in to no other possibilities except system shutdown.

1.1.3. Fault

A fault in a dynamical system is a deviation of the system structure or system parameters from the nominal situation(Blanke et al. 2006).

After the occurrence of a fault in plant/system, the presence of fault is determined first. Further, the fault location is determined and then the severity/magnitude of fault is obtained. The controller structure or its parameters are then modified in order to control the plant/system in presence of a fault.

1.2. A General Approach

In brief, the following steps are followed for Fault Detection, Diagnosis and Fault tolerant control of any plant/system as shown in Fig.1.

- Fault Detection : To determine the occurrence of a fault → Yes/No?
- Fault Isolation : To determine the location of a fault → Where?
- Fault Identification/Estimation : To determine the condition/magnitude of fault → How much?

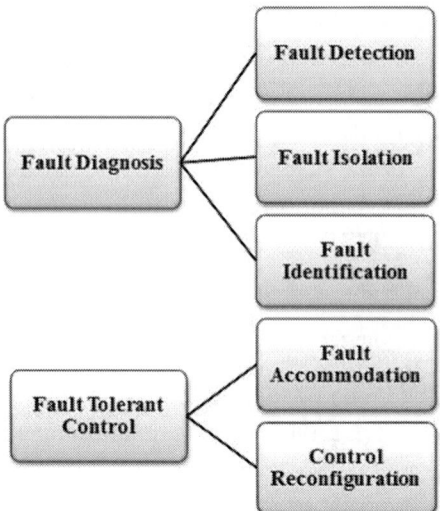

Figure 1. Steps of Fault Diagnosis and methods of Fault tolerant control.

- Fault Tolerant Control : To design a control strategy to mitigate the effect of estimated fault
  - Fault Accomodation : Update the parameters of the controller
  - Control Reconfiguration : In cases where Fault Accomodation is not possible, an alternate I/O pair is selected and a new control law is designed.

1.3. Types of Faults

In literature, faults are classified in several ways. The three forms of classification based on their location, its affect on signals and time profile of faults respectively, are as follows:

- Sensor faults, Actuator faults, Plant/System Faults
- Additive faults, Multiplicative Faults
- Abrupt faults, Incipient faults, Intermittent faults

Errors can occur in a sensor or actuator of a process which can be a constant bias or out of range conditions. The physical structure of the process can also

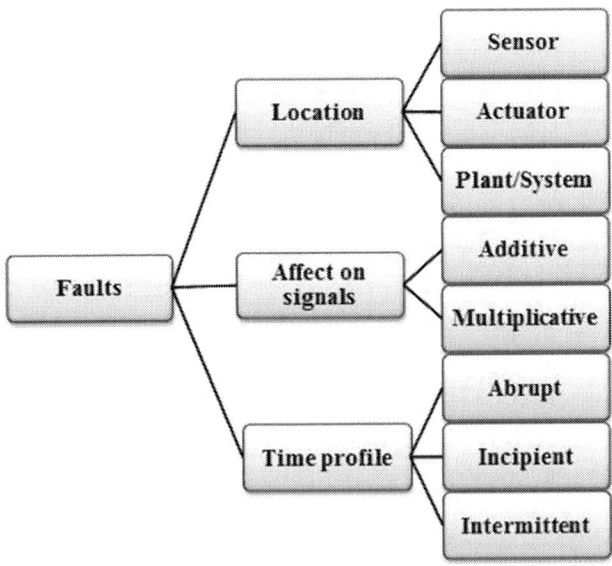

Figure 2. Classification of faults.

change resulting in a fault. The structural faults can be leaking pipe, change in nominal parameter value, improper functioning of the controller, stuck valve, etc. Additive faults appear as additive terms in the process equations and can be seen as unknown functions of time multiplying known matrices in system representation. On the other hand, multiplicative faults appear as terms that multiply with the inputs or outputs and can be seen as known functions of time multiplying unknown matrices. A fault that occurs suddenly can be termed as abrupt fault while the one which develops gradually over a period of time is incipient fault. Intermittent faults occur frequently for short periods before returning to normal operation.

1.4. Fault Tolerant Control

The fault tolerant control systems (FTCS) are also addressed as self repairing, self healing, reconfigurable and reliable control systems across literature (Venkatasubramanian et al. 2003a; Venkatasubramanian, Rengaswamy, and Kavuri 2003; Venkatasubramanian et al. 2003b; Zhang and Jiang 2008). The fault tolerant control systems(FTCS) can be classified into two types:

- Passive FTCS (PFTCS)
- Active FTCS (AFTCS)

1.5. Passive FTCS

In PFTCS, controllers are fixed and are designed to be robust against a class of presumed faults. This approach needs neither FDD schemes nor controller reconfiguration. This is because the control law is fixed and shall remain the same. It is more of a general control law that is determined to provide satisfactory performance for a class of known apriori faults. Thus it has limited fault-tolerant capabilities. The PFTCS are also known as reliable control systems or control systems with integrity (Zhang and Jiang 2008). Faults are sometimes referred to as uncertainties while Passive FTC (Fault tolerant control) approach is referred to as Robust control in control literature. Initially, PFTCS approach relied on using multiple controllers to achieve a reliable control system. The problem of passive FTC for actuator failures is studied by using H_2 optimization as well as in some cases by redundant components.

1.6. Active FTCS

In case of AFTCS, the control laws are not fixed as in case of PFTCS. The control laws are pre-computed or synthesized online for apriori faults. These control laws are updated according to faults and the controller compensates for the impact of the faults in order to achieve an acceptable performance of system operation. In rare cases, the degraded system performance may be accepted.

1.7. Rationale

The fault diagnosis as well as fault tolerant control methods have been well developed for linear or non-linear systems with various system representations. A growing interest is seen in research on the identification and control of dynamic systems described by DAE's during the past few years(Vemuri, Polycarpou, and Ciric 2001). However less attention is paid regarding fault diagnosis and fault tolerant control of systems described with Differential Algebraic equations. Since many complex systems like industrial chemical processes, mechanical systems, electrical systems etc., can naturally be modeled in descriptor form or expressed as Differential-Algebraic equations, it is thereby important to design fault diagnosis and fault tolerant schemes for the same. In the following section, we discuss Descriptor systems and their properties.

1.8. Descriptor Systems

A large class of engineering systems like process systems, mechanical systems, electrical systems, mechatronics, vehicle dynamics, water distribution network system, thermodynamic systems, etc. lead to DAE model equations (descriptor form). Descriptor systems are also referred to as implicit systems, singular systems, generalized state-space systems, semi-state systems and differential algebraic equation (DAE) systems (Gao and Ho 2006; Müller 2000; Nikoukhah et al. 1992). Please note that the terms DAE and decriptor systems are used interchangeably throughout the paper. The descriptor systems arise from dynamics of the process and laws of conservation, balance equations, constraints described by differential equations and algebraic equations respectively. For a long time, DAE's were simply considered to be implicit form of ODE's (Ordinary Differential Equations). However, in the early 1980s, the numerical integration methods on certain DAE's failed even after repeated attempts which initiated research in the field of DAE systems(Polycarpou, Vemuri, and Ciric 1997).

For system modeling the descriptor system approach has many advantages. It is a natural way to model process dynamics and provides more information about the physical behaviour of the system. Besides that, the interpretation of results obtained is simpler than in case of state space models which give an abstract description(Polycarpou, Vemuri, and Ciric 1997). A system expressed in the form of differential algebraic equations consists of differential equations as well as algebraic equations. Consider the system which can be expressed as:

$$F(\dot{x}, x, t) = 0 \qquad (1)$$

where x is a vector of dependent variables $x(t) = x_1(t), x_2(t), x_3(t), \ldots, x_n(t)$ and the system has equations $F = (F_1, F_2, F_3, \ldots, F_n)$. If the matrix $\frac{\partial F}{\partial \dot{x}}$ is singular (noninvertible), the system expressed as eq.1 is called a differential algebraic system of equations. They are alternatively known as descriptor systems. If the matrix $\frac{\partial F}{\partial \dot{x}}$ is non-singular then the system 1 can be transformed into ordinary differential equations of the form $\dot{x} = f(x, t)$. An index of a DAE is defined as the number of differentiations required of DAEs to get a system of

ODEs. A linear time invariant descriptor system can be expressed as,

$$E\dot{x}(t) = Ax(t) + Bu(t)$$
$$y(t) = Cx(t) + Du(t) \qquad (2)$$

where $A_{nxn}, B_{nxp}, C_{qxn}$ and D_{qxp} are system, input, output and feedthrough matrices respectively. The pair (E, A) is regular if

$$det(\lambda E - A) \neq 0 ; \quad \lambda \in C \qquad (3)$$

where C denotes the complex plane. This ensures that the system is solvable and possesses a unique solution for any given consistent value. The pair (E, A) is internally stable provided that,

$$rank(\lambda E - A) = n ; \quad \forall \lambda \in C_+ \qquad (4)$$

where C_+ denotes the closed and right-half plane. The pair is internally proper (also called impulse-free or causal) if,

$$rank \begin{bmatrix} E & 0 \\ A & E \end{bmatrix} = n + rank(E) \qquad (5)$$

or equivalently;

$$deg(det(\lambda E - A)) = rank(E) \qquad (6)$$

The triple (E,A,C) is impulse observable if

$$rank \begin{bmatrix} E & 0 \\ A & E \\ C & 0 \end{bmatrix} = n + rank(E) \qquad (7)$$

while system is considered finite detectable if,

$$rank \begin{bmatrix} \lambda E - A \\ C \end{bmatrix} = n ; \quad \forall \lambda \in C_+ \qquad (8)$$

Descriptor systems are also referred to as implicit systems, singular systems, generalized state-space systems, semi-state systems and differential algebraic equation (DAE) systems(Gao and Ho 2006).

1.9. Fault Modeling

Extending the model to include the faults is a widely adopted method of fault modeling in literature. Consider the descriptor system model of a linear time-invariant system described as eqns.(1-2). Any fault in sensors or actuators causes additive faults in the system. The descriptor system model with additive faults can then be given by

$$\begin{aligned} E\dot{x}(t) &= Ax(t) + Bu(t) + Fp(t) \\ y(t) &= Cx(t) + Du(t) + q(t) \end{aligned} \quad (9)$$

where $p(t)$ is actuator fault, $q(t)$ is sensor fault and F is the fault entry vector. Faults in process or structure of the plant causes a change in the model parameters resulting into multiplicative faults.

$$\begin{aligned} E\dot{x}(t) &= (A + \Delta A)x(t) + (B + \Delta B)u(t) \\ y(t) &= (C + \Delta C)x(t) + (D + \Delta D)u(t) \end{aligned} \quad (10)$$

where ΔA, ΔB, ΔC are changes in matrices because of faults which are in effect multiplied by states and the input to the system.

There are other methods of fault modeling wherein the state space model contains time dependent and exponential terms to model time profile of a fault. This allows to model abrupt or incipient faults. For example, an incipient and an abrupt fault can be modeled as(Vemuri, Polycarpou, and Ciric 2001),

$$\beta_i(t - T_0) = \begin{cases} = 0, & t < T_0 \\ = 1 - e^{-\alpha_i(t-T_0)}, & t \geq T_0 \end{cases} \quad (11)$$

where $\beta_i : R \to R$ is a function representing time profile of the fault; α_i represents fault evolution rate, T_0 is the time of fault occurence. A small value of α_i represents slowly developing fault. The same function which has a large value of α_i characterizes an abrupt fault. As shown in Fig.3, an abrupt fault of -1 magnitude and an incipient fault expressed in eq.11 with $T_0 = 250s$ and different values of α is shown. The fault modeling is done for the purpose of testing the efficacy of the algorithms during simulation. However, in real time implementation the faulty model can be considered to be replaced by physical setup from which data is acquired and fault detection, diagnosis and fault tolerant control strategy can be implemented.

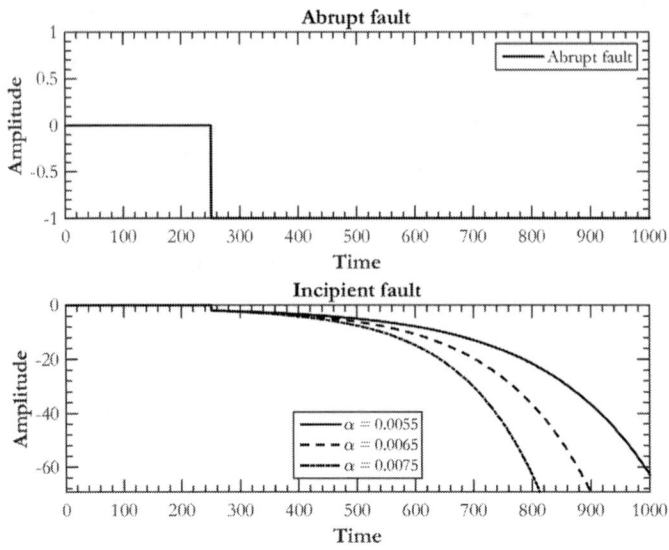

Figure 3. Time Profile : Abrupt & Incipient faults.

2. Fault Detection and Diagnosis

In general, the fault detection methods are classified into three broad categories as quantitative model-based methods, qualitative model-based methods and process history based methods (Venkatasubramanian et al. 2003a; Venkatasubramanian, Rengaswamy, and Kavuri 2003; Venkatasubramanian et al. 2003b). The methods can also be classified into : Data based methods and Model based methods(Miljković 2011). The fault diagnosis provides more information that includes fault location and fault magnitude. The Kalman filter based methods which rely on system models in descriptor form are discussed here.

2.1. Fault Detection

Model based fault detection techniques include output observers, state observers, parameter estimation and neural networks. The first step for observer based fault detection is to determine the estimated output. The ouput observer based approaches determine the estimated ouput by design of a suitable observer. In contrast, the state observer method obtains estimated state vector of the system by design of a suitable observer which along with measurement model determines estimated output. Once the predicted output is determined, the difference between the measured output and estimated output facilitates generation of residuals. The residual may also be passed through a filter or mapped using a residual evaluation function. The mapped or filtered signal so obtained is monitored for any threshold violations (pre-determined during fault free conditions) which indicates the occurrence of a fault.

2.1.1. Kalman Filter Based

The observer based approach for fault detection is discussed for systems in descriptor form (Vemuri, Polycarpou, and Ciric 2001; Yeu and Kawaji 2002; Hamdi et al. 2012; Zimmer and Meier 1997; Darouach and Boutayeb 1995; Bokor and Szabó 2009; Hamdi et al. 2010; Wang and Liang 2019). Majority of the works for residual generation for descriptor systems concerns residual generation by observer appraoach(Hamdi et al. 2010). As discussed earlier, the residuals can be generated from predicted outputs. The observer based methods for descriptor systems are discussed here keeping the fact in mind that the residuals generated with the information of predicted outputs can be compared with thresholds determined from nominal operating conditions for fault detection. It should be noted that the discussion is restricted to Kalman filter and its variants for descriptor systems.

To convert a DAE system into ODE system can seem the most appropriate solution as shown earlier, However, the observer based approaches if applied to the converted (ODE) system shall introduce problems such as

- inconsistent initialization of ode solvers since it shall not necessarily satisfy algebraic constraints.

- may result into system model consisting only differential states and lead towards inability to use measurement information which can be a function of algebraic states.

Hence it would deem to be appropriate to apply filters directly to descriptor systems.

The Kalman filters and its modifications are investigated for state estimation of linear and non-linear descriptor systems (Nikoukhah et al. 1992; Chisci and Zappa 1992; Becerra, Roberts, and Griffiths 2001; Kumar Mandela et al. 2010; Darouach, Boutayeb, and Zasadzinski 1997; Puranik et al. 2012; Pacharu,

Gudi, and Patwardhan 2012; Gupta and Hauser 2007; Haßkerl et al. 2016; Patel et al. 2019). Some works (Becerra, Roberts, and Griffiths 2001; Mandela, Rengaswamy, and Narasimhan 2001; Kumar Mandela et al. 2010; Puranik et al. 2012; Patel et al. 2019) demonstrate state estimation for non-linear descriptor sytems based on variants of Kalman filter by utilizing the system given explicitly. In those studies, the Kalman filters for descriptor system propagate the error and determine the differential states. The obtained differential states are further used to deduce algebraic states by solving constraints unlike Kalman filters applied to ODE system. However few works explore the application of such filters for fault detection (Wang, Shen, and Zhang 2012; Wang and Liang 2019).

The Kalman filters have been studied for linear discrete time descriptor systems(Chisci and Zappa 1992; Nikoukhah et al. 1992). In (Chisci and Zappa 1992), the proposed Kalman filter variant using a square root based approach can be applied to non-descriptor and descriptor systems with minor changes. A 3 block form of Kalman filter is obtained in (Nikoukhah et al. 1992) based on a maximum likelihood approach. The system expressed in descriptor form is considered to have a set of possibly noisy constraints on the behaviour of descriptor state x. Further, by means of Lemmas (Lemma 2.3 and 2.4 in Nikoukhah et al. 1992) it has been shown that the estimates obtained for the system considered under study is same as the estimates for a system described in general form as given by eq.2. The filtered estimate $\hat{x}$ is defined as the maximum likelihood estimate of x for the system under consideration along with the knowledge of initial conditions.

In an earlier work(Nikoukhah et al. 1992), a Kalman filter in 3 block form based on maximum likelihood estimation is derived for descriptor systems. The algorithm derived in that work is presented here in a diluted manner. Consider a LDS described in the form given as

$$Ex_{k+1} = Ax_k + Bu_k + \mu_k$$
$$y_{k+1} = Cx_{k+1} + v_k$$
(12)

For a system to be estimable, it is shown by (Nikoukhah et al. 1992) that the matrix $\begin{bmatrix} E \\ C \end{bmatrix}$ should have full column rank. From apriori knowledge of initial conditions and system expressed by eq.12, the maximum likelihood estimate of x can be obtained as $\hat{x}$. It is assumed that the initial state is Gaussian with known mean and variance. Given the filtered estimate $\hat{x}$ and eq.12 the following

equation is deduced
$$Y_k = Fx_{k+1} + W_k \tag{13}$$
where,
$$Y_k = \begin{bmatrix} A\hat{x}_k + Bu_k \\ y_{k+1} \end{bmatrix}, F = \begin{bmatrix} E \\ C \end{bmatrix}, W_k = \begin{bmatrix} w_k \\ v_k \end{bmatrix}$$
and
$$w_k = -A\tilde{x}_k - \mu_k, \tilde{x}_k = x_k - \hat{x}_k$$

In addition it has been shown that the systems expressed by eq.12 and eq.13 will result into same optimal estimates which implicitly establishes that both systems are equivalent. Further, the maximum likelihood estimate of x_{k+1} for system of eq.13 is given as

$$\hat{x}_{k+1} = (F^T \chi^{-1} F)^{-1} F^T \chi^{-1} Y_k \tag{14}$$

and the error covariance matrix is updated as

$$\hat{P}_{k+1} = (F^T \chi^{-1} F)^{-1} \tag{15}$$

where
$$\chi = \begin{bmatrix} AP_k A^T + Q & 0 \\ 0 & R \end{bmatrix}$$

The variance matrices Q and R are positive definite. Leveraging on the estimation approach discussed above, the fault detection is performed by augmenting descriptor system states with fault vector for attitude sensors in satellites(Wang, Shen, and Zhang 2012). This implies that such an approach as shown in (Wang, Shen, and Zhang 2012) would contribute to fault isolation and estimation simultaneously. However, the system under study seems to be simplified on account of the consideration of sensor faults only.

The Extended Kalman filter applied to nonlinear DAE systems in (Becerra, Roberts, and Griffiths 2001) is futher developed in (Kumar Mandela et al. 2010). In (Becerra, Roberts, and Griffiths 2001), the application of Extended Kalman Filter (EKF) to systems described by Non-Linear Differential Algebraic equations for state estimation is demonstrated. Consider the following DAE system,

$$\hat{x}_{k+1} = \hat{x}_k + \int_{k\Delta t}^{(k+1)\Delta t} f(x(t), z(t))dt + w_{k+1}$$
$$g(x(t), z(t)) = 0, \quad k\Delta T \le t \le (k+1)\Delta t \tag{16}$$

$$y_{k+1} = h(x_{k+1}) + v_{k+1} \tag{17}$$

where $x(t)$ and $z(t)$ are state variables and algebraic variables respectively while w_k and v_k are Gaussian white noise processes. The linearized system is obtained as,

$$\dot{x} = Ax \tag{18}$$

where $A = (J_1 - J_2 J_4^{-1} J_3)$ and $\begin{bmatrix} J_1 & J_2 \\ J_3 & J_4 \end{bmatrix} = \begin{bmatrix} \frac{\partial f}{\partial x} & \frac{\partial f}{\partial z} \\ \frac{\partial g}{\partial x} & \frac{\partial g}{\partial z} \end{bmatrix}$. The covariance error matrix of differential states is now obtained as in case of EKF

$$P_{k+1|k} = \Phi_k P_{k|k} \Phi_k^T + Q_{k+1} \tag{19}$$

where $\Phi_k = e^{(A\Delta t)}$. The estimates are then updated by computing the Kalman gain as

$$K_{k+1} = P_{k+1|k} H_{k+1}^T (H_{k+1} P_{k+1|k} H_{k+1}^T + R_{k+1})^{-1} \tag{20}$$

where $H_{k+1} = \frac{\partial h}{\partial x}$, h is the measurement model.

$$\hat{x}_{k+1|k+1} = \hat{x}_{k+1|k} + K_{k+1}(y_{k+1} - h(\hat{x}_{k+1|k})) \tag{21}$$

The error covariance matrix is propagated as

$$P_{k+1|k+1} = (I - K_{k+1} H_{k+1}) P_{k+1|k} \tag{22}$$

Finally, the updated estimates of the algebraic states are obtained by solving

$$g(\hat{x}_{k+1|k+1}, \hat{z}_{k+1|k+1}) = 0 \tag{23}$$

However the method proposed assumes that the measurements are function of differential states only.

The above mentioned limitation was overcome by method proposed in (Kumar Mandela et al. 2010). The linearized model for both differential and algebraic states is derived which is used for covariance error propagation of the states in augmented form. In that, the linearization of Non-linear DAE system is obtained in the form as follows,

$$\begin{aligned} \dot{x} &= J_1 x + J_2 z \\ 0 &= J_3 x + J_4 z \end{aligned} \tag{24}$$

where J_1, J_2, J_3, J_4 are the same as computed earlier in eqn.18. The linearized algebraic equation is differentiated and used to augment the system.

$$\dot{z} = -J_4^{-1} J_3 \dot{x} \quad (25)$$

$$\begin{bmatrix} \dot{x} \\ \dot{z} \end{bmatrix} = \begin{bmatrix} J_1 & J_2 \\ -J_4^{-1} J_3 J_1 & -J_4^{-1} J_3 J_2 \end{bmatrix} \begin{bmatrix} x \\ z \end{bmatrix} \quad (26)$$

Thus, the augmented form can be expressed as,

$$\dot{x}_{aug} = A_{aug} x_{aug} \quad (27)$$

where $x_{aug} = \begin{bmatrix} x \\ z \end{bmatrix}$ and $A_{aug} = \begin{bmatrix} J_1 & J_2 \\ -J_4^{-1} J_3 J_1 & -J_4^{-1} J_3 J_2 \end{bmatrix}$. The state transition matrix is then obtained as,

$$\phi = e^{A_{aug} \Delta t} \quad (28)$$

Further a similar procedure as explained earlier of (Becerra, Roberts, and Griffiths 2001) is followed. Since the system is augmented, the state transition matrix also contains the information of how algebraic states will propagate besides differential states. Thus the error covariance matrix is also updated in that manner. It is to be noted that the measurement model is a function of both differential and algebraic states in this case. Besides that, an approach to include unscented transformation for Unscented Kalman filter and recursive nonlinear dynamic data reconciliation is also evaluated for estimation of DAE systems. In addition, a heuristic EKF(Huang, Reklaitis, and Venkatasubramanian 2003) is proposed and applied for fault detection for a fluid catalytic cracking unit. The heuristic is expressed in form of pseudo-measurements is used to reduce the linearization errors caused by implementation of EKF for nonlinear descriptor systems.

Recently, the use of Ensemble Kalman filter with some modifications for state estimation of systems described by DAE's is demonstrated (Puranik et al. 2012,Patel et al. 2019). The non-linear DAE system under consideration can be given as,

$$\dot{x} = f(x, z, u)$$
$$g(x, z, u) = 0 \quad (29)$$
$$y(t) = h(x, z)$$

In discrete form with sampling time T and considering process noise and measurement noise, the system can be given as follows,

$$x_{k+1} = x_k + \int_{kT}^{(k+1)T} f(x(\tau), z(\tau), \tau) d\tau + w_k$$
$$x_{k+1} = F(x_k, z_k, u_k) + w_k \quad (30)$$
$$g(x(\tau), z(\tau), \tau) = 0$$

and the measurement given as

$$y_k = h(x_k, z_k) + v_k \quad (31)$$

where w_k and v_k are white noise with zero mean and known distribution while they are mutually uncorrelated independent random variables. In EnKF for such systems, at initial time $t = 0$, first, N particles of each of the differential states, $\hat{x}_0^i$, $(i = 1, 2, \ldots N)$ are drawn from a known distribution with mean $\hat{x}_0$ for which corresponding algebraic states are determined using algebraic constraint equations. The particles are then propagated by the DAE model and process noise is added to the resultant states, which can be given as,

$$\hat{x}_{k+1|k}^i = x_{k+1|k}^i + w_{d|k}^i \quad (32)$$

The algebraic states are calculated using algebraic constraints to ensure that they remain consistent. The output is then given by

$$\hat{y}_{k+1|k}^i = h(x_{k+1|k}^i, x_{k+1|k}^i) + v_{k+1}^i \quad (33)$$

The mean of differential states, algebraic states and predicted output are obtained and a constrained optimization problem is solved for every ensemble of differential and algebraic states. The objective function is given as(Puranik et al. 2012),

$$(\hat{x}_{k+1|k}, \hat{z}_{k+1|k}) = \min_{x_{k+1}, z_{k+1}} \epsilon_{k+1}^T P_{k+1|k}^{-1} \epsilon_{k+1} + e_{k+1}^T R^{-1} e_{k+1} \quad (34)$$

where

$$P_{k+1|k} = \frac{1}{N_e - 1} \sum_{i=1}^{N_e} \left\{ \begin{bmatrix} \hat{x}_{k+1|k}^i \\ \hat{z}_{k+1|k}^i \end{bmatrix} - \begin{bmatrix} \bar{x}_{k+1|k} \\ \bar{z}_{k+1|k} \end{bmatrix} \right\} \left\{ \begin{bmatrix} \hat{x}_{k+1|k}^i \\ \hat{z}_{k+1|k}^i \end{bmatrix} - \begin{bmatrix} \bar{x}_{k+1|k} \\ \bar{z}_{k+1|k} \end{bmatrix} \right\}^T$$
(35)

and

$$\epsilon_{k+1} = \begin{bmatrix} x_{k+1} \\ z_{k+1} \end{bmatrix} - \begin{bmatrix} \hat{x}_{k+1|k} \\ \hat{z}_{k+1|k} \end{bmatrix} \quad (36)$$

$$e_{k+1} = y_{k+1} - \hat{y}_{k+1|k}$$

It is important to note that Ensemble Kalman filter (EnKF) is independent of the index of DAE system. It also works well in presence of non-Gaussian noise. It has been demonstrated that the performance of EnKF is better than Extended Kalman filters as well as Unscented Kalman filters(UKF). However this comes at the cost of computation time. The computational cost is directly proportional to the number of ensemble members in Ensemble Kalman filter because it solves a constrained optimization problem for each ensemble member during every iteration. The EnKF draws the initial states from a Gaussian distribution and propagated through as shown earlier. In contrast, the UKF relies on statistical linearization of the system using sigma points for approximation of mean and covariance through nonlinear transformations. A comparison of UKF, EKF and EnKF is also shown by simulation studies for state estimation of Packed Bed Reactors(Pacharu, Gudi, and Patwardhan 2012). In that, the Ensemble Kalman filter uses an ensemble of 45 members and the performance turns out to be better than Unscented Kalman filter and Extended Kalman filter. Also, the Unscented Kalman performs better than Extended Kalman filter.

In (Kumar Mandela et al. 2010), the Unscented Kalman filter formulation for descriptor systems is proposed. Let the filtered differential state estimates at k^{th} instant be $\hat{x}_{k|k}$. 2n+1 sigma points $\hat{X}_{k|k,i}$ with associated weights are chosen symmetrically about $\hat{x}_{k|k}$ where n is the dimension of the state.

$$\hat{X}_{k|k,0} = \hat{x}_{k|k}$$
$$W_0 = \frac{K}{n+K} \quad (37)$$

$$\hat{X}_{k|k,i} = \hat{x}_{k|k} + (\sqrt{(n+K)P_{k|k}})_i$$
$$W_i = \frac{1}{2(n+K)} \quad (38)$$

$$\hat{X}_{k|k,i+n} = \hat{x}_{k|k} - (\sqrt{(n+K)P_{k|k}})_i$$
$$W_{i+n} = \frac{1}{2(n+K)} \quad (39)$$

$P_{k|k}$ is associated covariance matrix and $(\sqrt{P_{k|k}})_i$ is the ith column of the covariance matrix while K is the tuning parameter. The sum of weights W_i is unity. From $\hat{X}_{k|k,i}$ and algebraic constraint function g (similar to the one discussed earlier in eq.30) obtain $\hat{Z}_{k|k,i}$. These consistent sigma points are propagated using a DAE solver and $\hat{X}_{k+1|k,i}, \hat{Z}_{k+1|k,i}$ are obtained. The predicted differential state estimate $\hat{x}_{k+1|k}$ is obtained as

$$\hat{x}_{k+1|k} = \Sigma_{i=0}^{2n} W_i \hat{X}_{k+1|k,i} \quad (40)$$

Further,

$$P_{k+1|k}^{xx} = \Sigma_{i=0}^{2n} W_i (\hat{X}_{k+1|k,i} - \hat{x}_{k+1|k})(\hat{X}_{k+1|k,i} - \hat{x}_{k+1|k})^T + Q_{k+1} \quad (41)$$

The unscented sampling is performed next with $\hat{x}_{k+1|k}$ as mean and $P_{k+1|k}^{xx}$ as covariance matrix. From these unscented samples $\hat{X}_{k+1|k,i}$, using the algebraic equation $\hat{Z}_{k+1|k,i}$ is obtained. Further, $\hat{X}_{k+1|k,i}^{aug}$ is obtained by augmenting $\hat{X}_{k+1|k,i}$ with $\hat{Z}_{k+1|k,i}$ and then $\hat{x}_{k+1|k,i}^{aug}$ is calculated as done earlier in eq.40. Then, $Y_{k+1,i}$ is obtained by the measurement model using augmented states. Then the Kalman gain is computed as

$$K_{k+1} = (\Sigma_{i=0}^{2n} W_i (\hat{Y}_{k+1,i} - \hat{y}_{k+1})(\hat{Y}_{k+1,i} - \hat{y}_{k+1})^T + R_{k+1}) \\ (\Sigma_{i=0}^{2n} W_i (\hat{X}_{k+1,i}^{aug} - \hat{x}_{k+1|k}^{aug})(\hat{Y}_{k+1,i} - \hat{y}_{k+1})^T)^{-1} \quad (42)$$

Let the Kalman gain corresponding to differential states be expressed as K_{k+1}^d. Thus the differential states can be updated as

$$\hat{x}_{k+1|k+1} = \hat{x}_{k+1|k} + K_{k+1}^d (y_{k+1} - \hat{y}_{k+1}) \quad (43)$$

Given $\hat{x}_{k+1|k+1}$; $\hat{z}_{k+1|k+1}$ can be obtained by algebraic constraint equation as obtained earlier. Finally the error covariance matrix is updated in the last step as

$$P_{k+1|k+1} = P_{k+1|k} - K_{k+1}^d (\Sigma_{i=0}^{2n} W_i (\hat{Y}_{k+1,i} - \hat{y}_{k+1})(\hat{Y}_{k+1,i} - \hat{y}_{k+1})^T + R_{k+1})) K_{k+1}^{d^T} \quad (44)$$

The UKF filter so discussed above is modified and evaluated for fault detection by (Alkov and Weidemann 2013) for nonlinear descriptor systems. Three UKF's are implemented to generate robust residuals and determine the system operation. For a particular fault, the corresponding UKF generates residuals to determine its occurrence.

2.1.2. Residual Evaluation

For all observer based approaches including Kalman filters, a simple limit based fault detection strategy similar to that applied in (Foo, Zhang, and Vilathgamuwa 2013) can be adopted. Some of the simplest and computationally inexpensive residual evaluation methods that facilitate for fault detection are(*Model-based Fault Diagnosis Techniques*):

- Limit monitoring
  For a given signal r, the most primitive form of monitoring can be given as

$$r < r_{min} \text{ or } r > r_{max} \; ; \; fault\ detected$$
$$r_{min} \leq r \leq r_{max} \; ; \; fault\ free, \; nominal\ operation \quad (45)$$

where r_{min} and r_{max} denotes the minimum and maximum values of r during nominal system operation. The r_{min} and r_{max} are considered to be thresholds for such form of monitoring. Similarly, the trend analysis in the form of rate of change can be used to determine the limits and checked for threshold violations. The trend analysis i.e. rate of change in this case of signal r can be expressed as

$$\dot{r}(t) < \dot{r}(t)_{min} \text{ or } \dot{r}(t) > \dot{r}(t)_{max} \; ; \; fault\ detected$$
$$\dot{r}(t)_{min} \leq \dot{r}(t) \leq \dot{r}(t)_{max} \; ; \; fault\ free, \; nominal\ operation \quad (46)$$

where $\dot{r}_{min}$ and $\dot{r}_{max}$ are minimum and maximum values of $\dot{r}$. The rate of change of signal for discrete time case is given by

$$\Delta r = r(k) - r(k-1) \quad (47)$$

and the thresholds with minimum and maximum values of Δr can be determined as well and the fault occurrence can be detected similarly as eq.46 The peak value of $\dot{r}$ can be employed to redefine trend analysis.

$$\dot{r}(t)_{trendnorm} = \|\dot{r}(t)\| \quad (48)$$

. Similarly, we can obtain the redefined trend analysis for a discrete time case by using 47 and 48. The thresholds are determined apriori during fault free or nominal operating conditions for continuous and discrete time systems respectively as

$$\dot{r}(t)_{threshold} = \sup_{nominal} \|\dot{r}(t)_{trendnorm}\| \quad (49)$$

for $t \geq 0$

$$\Delta r_{threshold} = \sup_{nominal} \|\Delta r(k)_{trendnorm}\| \quad (50)$$

for $k \geq 0$
and the decision of fault occurence is taken when the residual evaluation function signal i.e. trend analysis signal in this case exceeds the thresholds determined by eq.49 & eq. 50

- **RMS (Root mean square) function**
  The root mean square roots itself from statistics and its applications. It can be defined as the square root of the mean of the squares of a set of signal magnitude. The RMS is a particular case of generalized mean with exponent as 2 and is alternatively known as quadratic mean. In case where the RMS value to be defined is for a continuous signal it can be defined as a function in terms of an integral of the squares of the instantaneous values. The RMS value measures the average energy of the signal over a certain time interval $(0, T)$ for continuous systems or for certain number of samples n, looking back into the past. Intuitively, it can be understood as a moving window/buffer that considers measurements from r_k to r_{k-n} where n is the number of signal samples. With regards to estimation theory, the RMS value of the signal is considered to be the measure of the imperfection of the fit of the estimator to the data. Thus, the RMS value determined for nominal system operation apriori can act as an indicator and serve for the purpose of how imperfectly the residual signal fits (in cases of faulty operation) as compared to the residual signal obtained during nominal operation. Thus, if for a certain nominal operation residual signal with r_{RMS} as RMS value; will act as a threshold.

  The RMS value is denoted by $\|.\|_{RMS}$. In case of discrete system for n samples of signal, the RMS value of the signal is given as

$$\|r(k)\|_{RMS} = \sqrt{\frac{1}{n}\left(r_k^2 + r_{k-1}^2 + r_{k-2}^2 \cdots + r_{k-n}^2\right)} \quad (51)$$

$$\|r(k)\|_{RMS} = \sqrt{\frac{1}{n}\sum_{i=1}^{n}\|r_{k-i}\|^2} \quad (52)$$

For a continuous system, the RMS value of the signal can be expressed as

$$\|r(t)\|_{RMS} = \sqrt{\frac{1}{T_2 - T_1} \int_{T_1}^{T_2} \|r(\tau)\|^2 d\tau} \qquad (53)$$

$$\|r(t)\|_{RMS} = \sqrt{\frac{1}{T} \int_{t}^{t+T} \|r(\tau)\|^2 d\tau} \qquad (54)$$

The solution to the fault detection problem can be posed as

$$\begin{aligned}\|r\|_{RMS} < \|r\|_{RMSmin} \;\; or \;\; \|r\|_{RMS} > \|r\|_{RMSmax} \;\;;\;\; fault\;detected \\ \|r\|_{RMSmin} \leq r \leq \|r\|_{RMSmax} \;\;;\;\; fault\;free\;,\;nominal\;operation\end{aligned} \qquad (55)$$

such that $\|r\|_{RMSmin}$ and $\|r\|_{RMSmax}$ are minimum and maximum value of $\|r\|_{RMS}$.

- Average function
The state estimation of systems sometimes may still be little noisy and hence the predicted output would be erratic as well. Besides that the measurement itself is corrupted by measurement noise. Thus, the residual signal will consist of some noise. In order to overcome such difficulty, the average function can be used. Intuitively, it can be understood as a moving window/buffer that considers measurements from r_k to r_{k-n} where n is the number of signal samples. For a continuous time case, the average function can be given as

$$\bar{r}(t) = \frac{1}{T} \int_{t}^{t+T} r(\tau) d\tau \qquad (56)$$

The average function for a discrete time signal can be expressed as

$$\bar{r}(k) = \frac{1}{n}(r_k + r_{k-1} + r_{k-2} + ... + r_{k-n}) \qquad (57)$$

$$\bar{r}(k) = \frac{1}{n} \sum_{i=1}^{n} r_{k-i} \qquad (58)$$

and for a continuous time case as

$$\bar{r}(t) = \frac{1}{T_2 - T_1} \int_{T_1}^{T_2} r(\tau)d\tau \qquad (59)$$

$$\bar{r}(t) = \frac{1}{T} \int_{t}^{t+T} r(\tau)d\tau \qquad (60)$$

The thresholds are then determined for discrete time case as

$$\bar{r}(k)_{threshold} = \sup_{nominal} \bar{r}(k) \qquad (61)$$

and continuous time case as

$$\bar{r}(t)_{threshold} = \sup_{nominal} \bar{r}(t) \qquad (62)$$

The decision regarding fault occurrence can then be given as

$$\begin{aligned}\bar{r}(k) > \bar{r}(k)_{threshold} &\ ;\ fault\ detected \\ \bar{r}(k) \leq \bar{r}(k)_{threshold} &\ ;\ fault\ free;\ nominal\ operation\end{aligned} \qquad (63)$$

and similarly

$$\begin{aligned}\bar{r}(t) > \bar{r}(t)_{threshold} &\ ;\ fault\ detected \\ \bar{r}(t) \leq \bar{r}(t)_{threshold} &\ ;\ fault\ free;\ nominal\ operation\end{aligned} \qquad (64)$$

- Peak function
  Consider that the residual signal $r \in R^{rn}$. The peak value function is defined for continuous time case as

$$r_{peak}(t) = \sqrt{\sum_{i=1}^{rn} r_i^2(t)} \qquad (65)$$

and for discrete time case as

$$r_{peak}(k) = \sqrt{\sum_{i=1}^{r_n} r_i^2(k)} \qquad (66)$$

The thresholds are then determined for discrete time case as

$$r_{peak}(k)_{threshold} = \sup_{nominal} r_{peak}(k) \qquad (67)$$

and continuous time case as

$$r_{peak}(t)_{threshold} = \sup_{nominal} r_{peak}(t) \qquad (68)$$

The decision regarding fault occurrence can then be given on the basis of it exceeding its thresholds. Some other statistical approaches*Model-based Fault Diagnosis Techniques* of residual evaluation for fault detection includes generalized likeliehood ratio, change in variance or chi square test. The likeliehood ratio test is featured here in a general formulation.

- Generalized Likelihood Ratio
  The generalized likelihood ratio (GLR) test determines the goodness of fit between two statistical models. Given a statistical model with parameter space Θ it can be hypothesized that the parameter θ is in specified subset Θ_0 of Θ. The alternative hypothesis is such that the parameter θ is in complement of Θ_0 denoted as Θ_{C0}. The likelihood ratio can be expressed as

$$GLR = -2\ln \frac{\sup_{\theta \in \Theta_0} L(\theta)}{\sup_{\theta \in \Theta} L(\theta)} \qquad (69)$$

As mentioned earlier in Section 1.2, the fault isolation procedure follows after "certain" detection of fault occurrence. The fault isolation approaches adopted across literature is outlined in the next section.

2.2. Fault Isolation

In general, the approach used for fault detection decides how to proceed with fault isolation for any system. Fault isolation deals with localization of occurred faults and deals with f_i of the fault vector f the decision on the presence of ith

fault ($f_i \neq 0$) or its absence ($f_i \neq 0$). Theoretically, the exact localization must be achieved regardless of faults occurring one at a time or several faults occur simultaneously. This makes fault isolation task significantly more challenging than fault detection.

2.2.1. Dedicated Observer Approach

The observer or filter of similar structure used for fault detection forms a basic element in the fault isolation scheme. A bank of observers or filters is used. In most cases, the number of observers used for fault isolation is equal to the number of faults to be detected. If the fault set F, has n faults, the isolation scheme will have n observers. The entire fault detection and isolation schema put together would have n+1 observers. The observers which are a part of fault isolation scheme are activated only after fault detection. All of these have similar structure but they are only sensitive to a particular fault. The observer which generates the residual in response to a particular fault facilitates for fault isolation. In an investigation based on UKF(Alkov and Weidemann 2013), the information of fault isolation is obtained in an implicit manner. Three conditions of system operation and their respective models are defined. A bank of UKF's is employed which determines the state of operation of the system. For a certain condition, the corresponding UKF utilizing the relevant system model is sensitive to that particular system operation.

2.2.2. Generalized Observer Approach

In certain cases, an exactly opposite approach is used. From the entire bank of observers, only one observer will not respond to a specific fault, while others would. This again forms a trivial solution for fault isolation.

2.2.3. Fault Isolation Filters

The central idea is to make each residual signal sensitive or insensitive to only a specific fault. A necessary and sufficient condition for two faults to be isolable by a given residual is that they do not belong to the same equivalence class(Staroswiecki and Comtet-Varga 2001). On the other hand, in a recent work(Wang and Liang 2019), an adaptive unscented Kalman filter is employed for sensor fault diagnosis, however in a multi model approach. An interesting analysis regarding fault isolation specification requirement for differential algebraic systems is also studied(Frisk, Krysander, and Åslund 2009). It determines minimal set of sensors to be added in order to achieve fault detectability and fault isolabilty. A set of minimal sensor sets are considered to be capable of fault detection if they have a non-null intersection with all fault identifiability sets. This uses existing analytical relationships and is applicable in cases where structural approaches shall fail. Consider fault $f_i \in F$, where $i = 1, 2, 3$ and F is the fault set. The isolability of the faults can proceed by posing the problem of isolation into sub-problems. To achieve fault isolation of the system considered, there are two sub-problems: 1) to isolate fault f_2, f_3 from f_1 2) to isolate fault f_2 from f_3. Solving these two sub-problems would be sufficient to achieve complete fault isolation. They present a formalized approach for fault isolation of systems described by linear differential algebraic equations. However such an approach might suffer from problems related to combinations of sensors because of the exhaustive nature of problem objective.

In general literature that considers the problem of fault isolation, isolating a single fault at a time is investigated. This arises from the fact that the probability of the occurrence of two or more faults at the same time is very low in physical systems.

2.3. Fault Estimation

The literature for fault estimation of descriptor systems in conjunction with Kalman filters is rare. In (Ali et al. 2014), the descriptor system Kalman filter is used to generate an innovation sequence. In absence of a fault or nominal operation the residual sequence is white noise. It is further shown that the innovation sequence can be used to estimate possible changes in the parameter vector. The implementation of the approach shown requires knowledge of the nominal parameter. Additionally, by appropriately filtering the input-output data, the problem is shown to be equivalent as a classical linear regression problem. In contrast, some approaches tend to be model independent which are based on neural networks. In (Vemuri, Polycarpou, and Ciric 2001), the problem of fault estimation for nonlinear descriptor systems is addressed by sigmoidal neural networks as online approximators. A general nonlinear descriptor system is considered to have modeling errors that are unstructured and bounded which is subjected to abrupt or incipient faults, and can be expressed as,

$$\dot{x} = F_x(x, z, u) + \Delta_x(x, z, u, t) + \beta_x(t - t_f)f_x(x, z, u)$$
$$0 = F_z(x, z, u) + \Delta_x(x, z, u, t) + \beta_z(t - t_f)f_z(x, z, u)$$
(70)

where $x \in R^{n-r}$, $z \in R^r$ and $u \in R^s$ are state variable vector, algebraic variable vector and input vector, respectively. Δ_x and Δ_y represent unknown modeling uncertainty with known bounds, f_x and f_z represent smooth vector fields while $\beta_x(t - t_f)$ and $\beta_z(t - t_f)$ represents the time profile of a fault which are in diagonal form. The incipient fault is represented by

$$\beta_i(t - t_f) = \begin{cases} = 0, & t < t_f \\ = 1 - e^{-\alpha_i(t - t_f)}, & t \geq t_f \end{cases}$$
(71)

while abrupt faults by

$$\beta_i(t - t_f) = \begin{cases} = 0, & t < t_f \\ = 1, & t \geq t_f \end{cases}$$
(72)

where $i \in (x, z)$ and t_f is time of fault occurrence. Consider the nonlinear estimation scheme in the form of

$$\dot{\hat{x}} = F_x(x, z, u) + \hat{f}_x(x, z, u; \hat{\theta}_x) + G\hat{x} - Gx$$
$$0 = F_z(x, z, u) + \hat{f}_z(x, z, u; \hat{\theta}_z)$$
(73)

where G is a positive definite diagonal matrix of dimension n-r and the elements of diagonal are the ith filter pole given as $-g_i$. The nonlinear approximators for state variables expressed as $\hat{f}_x(x, z, u; \hat{\theta}_x)$ and for algebraic variables expressed as $\hat{f}_z(x, z, u; \hat{\theta}_z)$ provide for estimation of the fault where (x, z, u) is input to the network, $\hat{\theta}$ is a vector of adjustable parameter weights and $\phi = \hat{f}(x, z, u; \hat{\theta})$ is network output. The parameters of the approximator are updated on the basis of state estimation error and constraint output estimation error given by $e_x = x - \hat{x}$ and $e_z = -\hat{y}$ respectively. Note that $\hat{y}$ is the estimated constrained output and since algebraic equations equate to zero, $-\hat{y}$ represents algebraic constraint estimation error. The parameter update is given by

$$\dot{\hat{\theta}}_i = \mathbb{P}\left\{\eta_i Z_i^T D\left[e_i, \frac{\Delta_i}{g_i}\right]\right\}$$
(74)

where $\mathbb{P}$ is the projection operator that restricts the parameter estimates, η_i is the learning rate, $Z_i = \frac{\partial \hat{f}_i(x, z, u; \hat{\theta})}{\partial \hat{\theta}_i}$, g_i is nonlinear estimator pole. The projection operator guarantees stability of the learning algorithm in presence of a fault. The observer design in this case is independent of the index of DAE system. But the method for determining the poles of G i.e. g_i is not formalized. The simulation results are obtained for a reactive flash system. However, the fault is introduced in the differential states and in no way affects the algebraic states

3. FAULT TOLERANT CONTROL

Many survey papers (Zhang and Jiang 2006; Patton 1997; Blanke, B, and Lunau 1997; Jiang 2005; Benosman 2010; Zhang and Jiang 2008; Blanke, Staroswiecki, and Wu 2001) exist in literature that discuss about fault tolerant control approaches in general. However, the fault tolerant control of descriptor systems has not been explored at all, specifically, the ones that leverage on Kalman filters and its variants. However, the investigations that can be utilized to advance within this domain are pointed out. The fault tolerant control for a class of Lipschitz nonlinear descriptor systems is studied(Gao and Ding 2007a). In that, the descriptor states are estimated by a robust state space observer and the estimated states are used along with LMI (Linear Matrix Inequality) technique for a fault tolerant control of descriptor system in presence of faults. The faults considered here are unbounded with the assumption that its qth derivative is bounded. The augmented system is used to design observer whose gains are determined to ensure that the steady state error dynamics are minimized. Since the augmented states already consist of fault as well as derivatives of faults, simultaneous estimation of states and faults is realized by the observer. Further an observer based controller is designed whose control law is in the form of $u = K\hat{x}$, where $\hat{x}$ is the estimated augmented state vector. The feedback gain K is determined by solving LMI. The control law realized through LMI provides closed loop stable solution in presence of fault. The LMI's resulting explicitly from closed loop stability conditions are not in strict LMI form. So in order to resolve computational issues arising because of non-strict LMI's, the closed loop stability conditions were modified in order to obtain strict LMI's. In (Marx et al. 2007), Robust Fault Tolerant control for a class of descriptor systems is put forth whose central idea is based on co-prime factorization and Youla parameterization. The approach is unique in the sense that it relies on Youla parameterization for fault diagnosis, nominal control and fault tolerant control. Fault specific filters are synthesized and appropriate corresponding filters are selected on the basis of directional residuals. The co-prime factorization facilitates to decompose the system and allows proper filters to be implemented. A residual signal obtained from Youla parameterization is filtered to diagnose the fault. It is important to note that Youla parameterization determines the set of all stabilizing controllers and can be categorized as a passive fault tolerant control approach. Such an approach can be extended to design a controller after fault detection. Recently, a Kalman filter-based fault-tolerant control (KFFTC) strategy for a doubly-fed induction generator (DFIG) under voltage and current sensor faults is evaluated(Xiahou and Wu 2018). It should be noted that this approach is implemented for conventional models. However, an extension of this approach can be considered to descriptor systems since it relies on Kalman filter framework. It implements six Kalman filters that perform state estimation in parallel which is utilized to calculate residuals. The faults are isolated and further the corresponding Kalman filter is utilized for control reconfiguration. The approaches so outlined above can be considered for Kalman filter based FTC of descriptor systems.

4. DESCRIPTOR SYSTEM: EXAMPLES AND SOLVERS

Modeling of physical systems or processes in descriptor form is a very natural way to model dynamics and behavior of system/process. The model provides physical insight within the system thus providing rich information about its physical behaviour. Several works Kumar Mandela et al. 2010; Pacharu, Gudi, and Patwardhan 2012; Haßkerl et al. 2016; Yeu, Kim, and Kawaji 2008; Darouach 1993; Purohit, Patwardhan, and Mahajani 2015; Boulkroune, Zasadzinski, and Darouach 2010; Rodrigues et al. 2014 consider different systems for study of descriptor systems. In this section we provide typical models and parameters of linear as well as non-linear descriptor/DAE systems. For a linear descriptor system expressed as in eqn. 2, the matrices are given as followsDarouach 1993:

$$E = \begin{bmatrix} 1 & 1 & 1 & 0 \\ 2 & 0 & -1 & 0 \\ 0 & 1 & 0 & 1 \end{bmatrix}, \quad A = \begin{bmatrix} 1 & 1 & 0 & 0.59 \\ 0 & -1 & 0 & 0.50 \\ 1 & 0 & 1 & 0.09 \end{bmatrix}$$
$$B = \begin{bmatrix} 1 & 1 \\ 2 & 0 \\ 1 & 2 \end{bmatrix}, \quad C = \begin{bmatrix} 1 & 0 & 0 & 1 \\ 0 & 1 & -0.5 & 0 \\ 0 & 0 & 1 & 1 \end{bmatrix}$$
(75)

Similar examples can be found in several works Darouach and Boutayeb 1995;

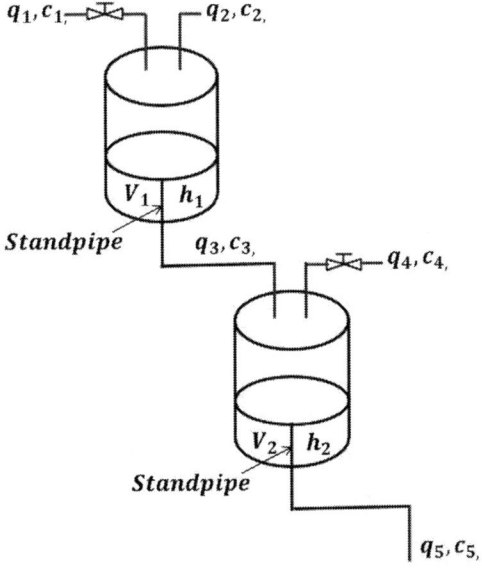

Figure 4. Chemical mixing process (Yeu, Kim, and Kawaji 2008).

Gao and Ding 2007b. In Yeu, Kim, and Kawaji 2008, a two-tank chemical mixing process is considered which is expressed in descriptor form as shown in Fig 4. It considers two inputs and three outputs. The heights, h_1 and h_2, of the liquid in the mixing tank are maintained constant. This introduces algebraic equations in the model by mass balance. For $q_i(t)$, $i \in [1, 5]$ and $c_j(t)$, $j \in [1, 5]$ which are flow rates and concentrations of input flow streams while V_l, $l \in [1, 2]$ are volumes of liquid in tank-1 and tank-2 respectively. The height in tank is maintained constant using a standpipe which removes any excess volume of liquid above height h_k, $k \in [1, 2]$. The inputs to the system are $u = [q_1(t), q_4(t)]$ controlled through valves. $q_2(t)$ is disturbance added to tank-1 while $q_3(t)$ & $q_5(t)$ are overflows from tank-1 and tank-2 respectively. The outputs of the system are $c_3(t)$, $q_3(t)$, $c_5(t)$ and $q_5(t)$. In (Boulkroune, Zasadzinski, and Darouach 2010), an electromechanical system consisting of DC motor with elastic coupling for load is considered. An electrochemical case study which models a galvanostatic charge process of a thin film nickel hydroxide electrode in the form of non-linear descriptor system is considered in several papers (Becerra, Roberts, and Griffiths 2001; Puranik et al. 2012; Mandela, Rengaswamy, and Narasimhan 2001;

Kumar Mandela et al. 2010; Patel et al. 2020). The state dynamics in form of ordinary differential equations is given as,

$$\frac{\rho V}{W} \frac{dx_1}{dt} = \frac{j_1}{F} \tag{76}$$

while the algebraic equation is given as,

$$j_1 + j_2 = I \tag{77}$$

where I is the applied current, x_1 is rate of change of mole fraction of nickel hydroxide and,

$$j_1 = i_{01}\left[2(1-x_1)exp\left(\frac{0.5F}{RT}(x_2 - \phi_{eq.1})\right) - 2x_1 exp\left(\frac{-0.5F}{RT}(x_2 - \phi_{eq.1})\right)\right]$$
$$j_2 = i_{02}\left[exp\left(\frac{F}{RT}(x_2 - \phi_{eq.2})\right) - exp\left(\frac{-F}{RT}(x_2 - \phi_{eq.2})\right)\right] \tag{78}$$

x_2 is the potential difference between the solid and liquid interface. The values of parameters used are: Faraday's constant $F = 96487 C/mol$, ideal gas constant $R = 8.314 J/molK$, temperature $T = 298.15K$, equilibrium potential of nickel reaction $\phi_{eq.1} = 0.420V$, equilibrium potential of oxygen reaction $\phi_{eq.2} = 0.303V$, density of nickel active material $r = 3.4g/cm^3$, molecular weight $W = 92.7g/mol$, effective length $V = 1 \times 10^{-5} cm$, applied current density on the nickel electrode $I = 1 \times 10^{-5} A/cm^2$, exchange current density of the nickel reaction $i_{01} = 1 \times 10^{-04} A/cm^2$ and exchange current density of the oxygen reaction $i_{02} = 1 \times 10^{-08} A/cm^2$. Besides these, a large scale DAE system of Reactive Distillation process for transesterification of dimethyl carbonate with ethanol is considered(Haßkerl et al. 2016). A reactive distillation column with hypothetical reaction kinetics and ideal vapour-liquid equilibrium is considered for study(Purohit, Patwardhan, and Mahajani 2015). The system has as many as 90 differential states that consist of liquid compositions of all stages and molar holdups. The system has 21 algebraic states which consists of temperature at all stages. Besides that, the system exhibits nonlinear input and output multiplicities and hence has only two steady state operating points. In (Rodrigues et al. 2014), A two phase flash system modeled in descriptor form is also considered as shown in Fig 5. Q, T_x, x_q are feed flow rate, feed temperature and volatile component mole-fraction in feed respectively. P, P_{ref} are pressure in flash tank and downstream pressure respectively while Q_L is liquid outflow rate. The system states are liquid mole fraction of volatile component

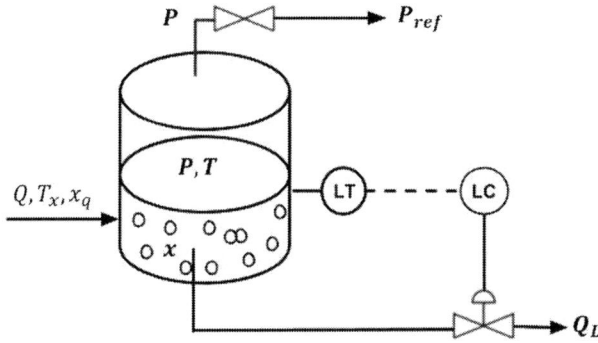

Figure 5. Flash tank system (Rodrigues et al. 2014).

$x_L(t)$, flash tank temperature $T(t)$, equilibrium mole fraction $x^*(t)$, pressure in flash tank $P(t)$ and liquid out-flow rate $Q_L(t)$. In that continuous isothermal reactor, a volatile component flashes from a binary diluted chemical and the level control is considered to be instantaneous. Since the system is considered to be isothermal implies that the gas and liquid phase are at the same temperature and accumulation of energy is negligible since $\gamma_l > \gamma_g$, where γ_l is mass fraction of liquid phase and γ_g is mass fraction of gaseous phase. Initially, for simulation of such systems, DASSL library was developed for index-1 problems. Runge Kutta solver and its modified versions like SDOP853 are available for solution of descriptor systems; that consist of an additional step of solving the constraints rising from algebraic equations. MATLAB allows for simulation of descriptor system state space models. Non-Linear descriptor systems can be simulated in MATLAB by defining a suitable "mass" matrix and ode15s function. The tools for analysis and design of descriptor systems have developed enormously in recent past (Müller 2000). Recently, the work done in (Varga 2017a) is a collection of MATLAB functions for the operation on and manipulation of rational transfer function matrices via their descriptor system realizations. They provide detailed information about syntax usage and functions. The algorithms implemented by the researchers are based on the computational procedures described in (Varga 2017b). A well maintained repository for the same is also available on BitBucket which can be used and directly implemented in MATLAB.

CONCLUSION

As an emerging domain fault diagnosis and fault tolerant control of systems in descriptor form/ described by differential algebraic equations has gained considerable attention. A review of the field of fault diagnosis and fault tolerant control of descriptor systems specifically focused on Kalman filters is presented here. The concepts and terms along with its alternative terms in the domain of fault tolerant control are discussed. Different approaches for classification of faults and fault tolerant control is presented in brief. The descriptor systems and its properties are presented in a comprehensive manner. The fault modeling of descriptor systems according to classification of faults is explained. The approaches of state estimation for descriptor system using Kalman filters and fault detection are discussed in detail. Sparse literature of fault isolation for descriptor systems is featured. The fault estimation with necessary details are furnished. The fault tolerant control of descriptor system is hardly explored. Lastly, the examples of various descriptor systems across literature are provided for reference. The literature for fault diagnosis and fault tolerant control specifically focused on descriptor systems is less. Particularly, these domains based on Kalman filter framework has a huge potential for future research. Additionally, it should be noted that the domain of fault detection of descriptor systems is well explored in comparison to that of fault diagnosis & fault tolerant control for descriptor systems. The investigations that can be potentially improvised for advancing research for descriptor systems are highlighted. The experimental results about real time implementation of fault diagnosis and fault tolerant control schema on physical systems are very sparse.

REFERENCES

Ali, Abdouramane Moussa et al. (2014). "Fault diagnosis for linear time-varying descriptor systems". In: *11th IFAC International Workshop on Adaptation and Learning in Control and Signal Processing,* Caen, France. pp.724-729.

Alkov, Ilja and Dirk Weidemann (2013). "Fault detection with unscented Kalman filter applied to nonlinear differential-algebraic systems". In: *2013 18th International Conference on Methods and Models in Automation and Robotics, MMAR 2013*, pp. 166–171. DOI: 10.1109/mmar.2013 6669900.

Becerra, V M, P D Roberts, and G W Griffiths (2001). "Applying the extended Kalman filter to systems described by nonlinear differential-algebraic equations". In: *PERGAMON Control Engineering Practice* 9, pp. 267–281. URL: www.elsevier.com/locate/conengprac.

Benosman, Mouhacine (2010). "A survey of some recent results on nonlinear fault tolerant control". In: *Mathematical Problems in Engineering* 2010. ISSN: 1024123X. DOI: 10.1155/2010/586169.

Blanke, M, S a B, and C P Lunau (1997). "Fault-tolerant control systems– A holistic view". In: *Control Engineering Practice* 5.5, pp. 693–702. ISSN: 0967-0661. DOI: http://dx.doi.org/10.1016/S0967-0661(97)00051-8.

Blanke, Mogens, Marcel Staroswiecki, and N Eva Wu (2001). "Concepts and Methods in Fault-tolerant Control". In: *American Control Conference*. URL: http://www.iau.dtu.dk/secretary/pdf/ACCmb30a4.pdf.

Blanke, Mogens et al. (2006). *Diagnosis and fault-tolerant control*. Vol. 691. ISBN: 9783540356523.

Bokor, József and Zoltán Szabó (2009). "Fault detection and isolation in nonlinear systems". In: *Annual Reviews in Control* 33.2, pp. 113–123. ISSN: 13675788. DOI: 10.1016/j.arcontrol.2009.09.001.

Boulkroune, B., M. Zasadzinski, and M. Darouach (2010). "Moving horizon state estimation for linear discrete-time singular systems". In: *IET Control Theory & Applications* 4.3, pp. 339–350. ISSN: 1751-8644. DOI: 10.1049/iet-cta.2008.0280. URL: http://digital-library.theiet.org/content/journals/10.1049/iet-cta.2008.0280.

Chisci, L. and G. Zappa (1992). "Square-root Kalman filtering of descriptor systems". In: *Systems and Control Letters* 19.4, pp. 325–334. ISSN: 01676911. DOI: 10.1016/0167-6911(92)90071-Y.

Darouach, M (1993). "State estimation of stochastic singular linear systems". In: *... Journal of Systems ...* 24.2, pp. 1–12. URL: http://www.tandfonline.com/doi/abs/10.1080/00207729308949493.

Darouach, M. and M. Boutayeb (1995). "Design of Observers for Descriptor Systems". In: *IEEE Transactions on Automatic Control* 40.7, pp. 1323–1327. ISSN: 15582523. DOI: 10.1109/9.400467.

Darouach, M., M. Boutayeb, and M. Zasadzinski (1997). "Kalman filtering for continuous descriptor systems". In: *Proceedings of the 1997 American Control Conference (Cat. No.97CH36041) 3.June, pp. 0–4. ISSN: 0743-1619. DOI: 10.1109/ACC.1997.611062.

Ding, Steven X. *Model-based Fault Diagnosis Techniques.* ISBN: 9781447147985.

Foo, Gilbert Hock Beng, Xinan Zhang, and D. M. Vilathgamuwa (2013). "A sensor fault detection and isolation method in interior permanent-magnet synchronous motor drives based on an extended kalman filter". In: *IEEE Transactions on Industrial Electronics* 60.8, pp. 3485–3495. ISSN: 02780046. DOI: 10.1109/TIE.2013.2244537. URL: https://www.researchgate.net/profile/Xinan{_}Zhang2/publication/260541837{_}A{_}Sensor{_}Fault{_}Detection{_}and{_}Isolation{_}Method{_}in{_}Interior{_}Permanent-Magnet{_}Synchronous{_}Motor{_}Drives{_}Based{_}on{_}an{_}Extended{_}Kalman{_}Filter/links/5553f7cc08ae980ca6086e34.pdf.

Frisk, Erik, Mattias Krysander, and Jan Åslund (2009). "Sensor placement for fault isolation in linear differential-algebraic systems". In: *Automatica* 45.2, pp. 364–371. ISSN: 00051098. DOI: 10.1016/j.automatica.2008.08.013. URL: http://dx.doi.org/10.1016/j.automatica.2008.08.013.

Gao, Z. and D.W.C. Ho (2006). "State/noise estimator for descriptor systems with application to sensor fault diagnosis". In: *IEEE Transactions on Signal Processing* 54.4, pp. 1316–1326. ISSN: 1053-587X VO - 54. DOI: 10.1109/TSP.2006.870579.

Gao, Zhiwei and Steven X. Ding (2007a). "Actuator fault robust estimation and fault-tolerant control for a class of nonlinear descriptor systems". In: *Automatica* 43.5, pp. 912–920. ISSN: 00051098. DOI: 10.1016/j.automatica.2006.11.018.

— (2007b). "Fault estimation and fault-tolerant control for descriptor systems via proportional, multiple-integral and derivative observer design". In: *IET Control Theory Appl* 16.5, pp. 1208–1218. ISSN: 14746670. DOI: 10.1049/iet-cta.

Gao, Zhiwei, Steven X. Ding, and Carlo Cecati (2015). "Real-time fault diagnosis and fault-tolerant control". In: *IEEE Transactions on Industrial Electronics* 62.6, pp. 3752–3756. ISSN: 02780046. DOI: 10.1109/TIE.2015.2417511.

Gertler, J (1991). "Analytical Redundancy Methods : Survey and Synthesis". In: *IFAC Proceedings Volumes* 24.6, pp. 9–21. ISSN: 1474-6670. DOI: 10.1016/S1474-6670(17)51119-2. URL: http://dx.doi.org/10.1016/S1474-6670(17)51119-2.

Gupta, Nachi and Raphael Hauser (2007). "Kalman Filtering with Equality and Inequality State Constraints". In: arXiv: 0709.2791. URL: http://arxiv.org/abs/0709.2791.

Hamdi, H. et al. (2012). "Fault detection and isolation in linear parameter-varying descriptor systems via proportional integral observer". In: *International Journal of Adaptive Control and Signal Processing* 26.3, pp. 224–240. ISSN: 08906327. DOI: 10.1002/acs.1260. arXiv: LPV{_}Xiaohang{_}2015. URL: http://onlinelibrary.wiley.com/doi/10.1002/acs.967/abstracthttp://doi.wiley.com/10.1002/acs.1260.

Hamdi, Habib et al. (2010). "Robust H infinity Fault Diagnosis for Multi-Model Descriptor Systems : A Multi-Objective Approach". In:

Haßkerl, Daniel et al. (2016). "Simulation Study of the Particle Filter and the EKF for State Estimation of a Large-scale DAE-system with Multi-rate Sampling". In: *IFAC-PapersOnLine* 49.7Haßkerl, D., Arshad, M., Hashemi, R., Subramanian, S., & Engell, S. (2016). Simulation Study of the Particle Filter and the EKF for State Estimation of a Large-scale DAE-system with Multi-rate Sampling. IFAC-PapersOnLine, 49(7), 490–495. https://doi.org/, pp. 490–495. ISSN: 24058963. DOI: 10.1016/j.ifacol.2016.07.390. URL: http://dx.doi.org/10.1016/j.ifacol.2016.07.390.

Huang, Yuanjie, G. V. Reklaitis, and Venkat Venkatasubramanian (2003). "A heuristic extended Kalman filter based estimator for fault identification in a fluid catalytic cracking unit". In: *Industrial and Engineering Chemistry Research* 42.14, pp. 3361–3371. ISSN: 08885885. DOI: 10.1021/ie010659t.

Isermann, Rolf (1984). "Process Fault Detection Based on Modeling and Estimation Methods - A Survey". In: *Automatica* 20.4, pp. 387–404. ISSN: 00051098. DOI: 10.1016/0005-1098(84)90098-0. URL: https://pdfs.semanticscholar.org/f8d3/5d3e5c3c9829b3a382b6f24c750eebf2b4d2.pdf.

Isermann, Rolf (2005). "Model-based fault-detection and diagnosis - Status and applications". In: *Annual Reviews in Control* 29.1, pp. 71–85. ISSN: 13675788. DOI: 10.1016/j.arcontrol.2004.12.002.

Jiang, J (2005). "Fault-tolerant Control Systems – An Introductory Overview." In: *Automatica SINCA* 31.1, pp. 161–174. ISSN: 02544156.

Kumar Mandela, Ravi et al. (2010). "Recursive state estimation techniques for nonlinear differential algebraic systems". In: *Chemical Engineering Science* 65.16, pp. 4548–4556. ISSN: 00092509. DOI: 10.1016/j.ces.2010.04.020. URL: http://dx.doi.org/10.1016/j.ces.2010.04.020.

Li, Linlin (2016). *Fault Detection and Fault-Tolerant Control for Nonlinear Systems*. ISBN: 978-3-658-13019-0. DOI: 10.1007/978-3-658-13020-6. URL: http://link.springer.com/10.1007/978-3-658-13020-6.

Mandela, R.K., Raghunathan Rengaswamy, and S. Narasimhan (2001). "Nonlinear State Estimation of Differential Algebraic Systems". In: *Nt.Ntnu.No.* ISSN: 1474-6670. DOI: 10.3182/20090712-4-TR-2008.00129. URL: http://www.nt.ntnu.no/users/skoge/prost/proceedings/adchem09/cd/abstract/31.pdf.

Marx, Benoît et al. (2007). "Robust Fault Tolerant Control for Descriptor Systems To cite this version : HAL Id : hal-00152242". In: *HAL*.

Miljković, Dubravko (2011). "Fault Detection Methods - A Literature Survey". In: *MIPRO, 2011 Proceedings of the 34th International Convention*. URL: https://bib.irb.hr/datoteka/515529.Fault{_}Detection{_}Methods{_}-{_}A{_}Literature{_}Survey.pdf.

Müller, P C (2000). "Descriptor systems: pros and cons of system modelling by differential-algebraic equations". In: *Mathematics and Computers in Simulation* 53.4-6, pp. 273–279. ISSN: 0378-4754. DOI: DOI: 10.1016/S0378-4754(00)00213-5. URL: http://www.sciencedirect.com/science/article/pii/S0378475400002135.

Nikoukhah, Ramine et al. (1992). "Kalman filtering and Riccati equations for descriptor systems". In: *IEEE Transactions on Automatic Control* 37.9, pp. 1325–1342. ISSN: 00189286. DOI: 10.1109/9.159570. URL: http://ieeexplore.ieee.org/document/159570/.

Noura, Hassan et al. (2009). *Fault-tolerant Control System Desing and Practial Applications*, pp. XXI, 233. ISBN: 9781848826526. DOI: 10.1007/978-1-84882-653-3.

Pacharu, Sreenivasa Rao, Ravindra Gudi, and Sachin Patwardhan (2012). *Advanced state estimation techniques for packed bed reactors*. Vol. 8. PART 1. IFAC, pp. 519–524. ISBN: 9783902823052. DOI: 10.3182/20120710-4-SG-2026.00185. URL: http://dx.doi.org/10.3182/20120710-4-SG-2026.00185.

Patel, T J et al. (2019). "Input Fault Detection using Ensemble Kalman Filter for NonLinear Descriptor Systems". In: *7th Nirma University International Conference on Engineering*.

Patel, Tigmanshu et al. (2020). "State estimation of NonLinear Descriptor Systems using Particle Swarm Optimization based Extended Kalman Filter". In: *European Control Conference*.

Patton, Ron J. (1997). "Fault-Tolerant Control: The 1997 Situation". In: *IFAC Proceedings Volumes* 30.18, pp. 1029–1051. ISSN: 14746670. DOI: 10.1016/S1474-6670(17)42536-5. URL: http://linkinghub.elsevier.com/retrieve/pii/S1474667017425365.

Polycarpou, Marios M, Arun T Vemuri, and Amy R Ciric (1997). "Nonlinear Fault Diagnosis in Differential Algebraic Systems by". In: *IFAC SafeProcess*.

Puranik, Yash et al. (2012). *An ensemble kalman filter for systems governed by differential algebraic equations (DAEs)*. Vol. 8. PART 1. IFAC, pp. 531–536. ISBN: 9783902823052. DOI: 10.3182/20120710-4-SG-2026.00167. URL: http://dx.doi.org/10.3182/20120710-4-SG-2026.00167.

Purohit, Jalesh L, Sachin C Patwardhan, and Sanjay M Mahajani (2015). "State Estimation of a Reactive Distillation System using Multi-rate DAE EKF". In: *Indian Control Conference*. March.

Rodrigues, M et al. (2014). "Observer-based fault tolerant control design for a class of LPV descriptor systems". In: *Journal of the Franklin Institute* 351.6, pp. 3104–3125. ISSN: 0016-0032. DOI: http://dx.doi.org/10.1016/j.jfranklin.2014.02.016. URL: http://www.sciencedirect.com/science/article/pii/S0016003214000520.

Staroswiecki, M. and G. Comtet-Varga (2001). "Analytical redundancy relations for fault detection and isolation in algebraic dynamic systems". In:

Automatica 37.5, pp. 687–699. ISSN: 00051098. DOI: 10.1016/S0005-1098(01)00005-X.

Varga, Andreas (2017a). "Descriptor System Tools (DSTOOLS) User's Guide". In: arXiv: 1707.07140. URL: http://arxiv.org/abs/1707.07140.

— (2017b). *Solving Fault Diagnosis Problems*. Vol. 84. ISBN: 978-3-319-51558-8. DOI: 10.1007/978-3-319-51559-5. URL: http://link.springer.com/10.1007/978-3-319-51559-5.

Vemuri, Arun T., Marios M. Polycarpou, and Amy R. Ciric (2001). "Fault diagnosis of differential-algebraic systems". In: *IEEE Transactions on Systems, Man, and Cybernetics Part A:Systems and Humans*. 31.2, pp. 143–152. ISSN: 10834427. DOI: 10.1109/3468.911372.

Venkatasubramanian, Venkat, Raghunathan Rengaswamy, and Surya N. Kavuri (2003). "A review of process fault detection and diagnosis part II: Qualitative models and search strategies". In: *Computers and Chemical Engineering* 27.3, pp. 313–326. ISSN: 00981354. DOI: 10.1016/S0098-1354(02)00161-8.

Venkatasubramanian, Venkat et al. (2003a). "A review of process fault detection and diagnosis Part I: Quantitative model-based methods". In: *Computers and Chemical Engineering*.

Venkatasubramanian, Venkat et al. (2003b). "A review of process fault detection and diagnosis Part III: Process history based methods". In: *Computers and Chemical Engineering*.

Wang, Mao and Tiantian Liang (2019). "Adaptive Kalman filtering for sensor fault estimation and isolation of satellite attitude control based on descriptor systems". In: *Transactions of the Institute of Measurement and Control* 41.6, pp. 1686–1698. ISSN: 01423312. DOI: 10.1177/0142331218787605.

Wang, Zhenhua, Yi Shen, and Xiaolei Zhang (2012). "Attitude sensor fault diagnosis based on Kalman filter of discrete-time descriptor system". In: *Journal of Systems Engineering and Electronics* 23.6, pp. 914–920. ISSN: 10044132. DOI: 10.1109/JSEE.2012.00112.

Xiahou, K. S. and Q. H. Wu (2018). "Fault-tolerant control of doubly-fed induction generators under voltage and current sensor faults". In: *International Journal of Electrical Power and Energy Systems* 98.September 2017, pp. 48–61. ISSN: 01420615. DOI: 10.1016/j.ijepes.2017.11.028.

Yeu, T.K. and S. Kawaji (2002). "Sliding mode observer based fault detection and isolation in descriptor systems". In: *2002 American Control Conference* 6, pp. 4543–4548. DOI: 10.1109/ACC.2002.1025367. URL: http://ieeexplore.ieee.org/lpdocs/epic03/wrapper.htm?arnumber=1025367.

Yeu, Tae-Kyeong, Hwan-Seong Kim, and Shigeyasu Kawaji (2008). "Fault Detection, Isolation and Reconstruction for Descriptor Systems". In: *Asian Journal of Control* 7.4, pp. 356–367. ISSN: 15618625. DOI: 10.1111/j.1934-6093.2005.tb00398.x. URL: http://doi.wiley.com/10.1111/j.1934-6093.2005.tb00398.x.

Zhang, Youmin and Jin Jiang (2006). *Issues on Integration of Fault Diagnosis and Reconfigurable Control in Active Fault-Tolerant Control Systems.* Vol. 39. 13. IFAC, pp. 1437–1448. ISBN: 9783902661142. DOI: 10.3182/20060829-4-CN-2909.00240. URL: http://linkinghub.elsevier.com/retrieve/pii/S147466701534622X.

— (2008). "Bibliographical review on reconfigurable fault-tolerant control systems". In: *Annual Reviews in Control* 32.2, pp. 229–252. ISSN: 13675788. DOI: 10.1016/j.arcontrol.2008.03.008.

Zimmer, Gerta and Jürgen Meier (1997). "On observing nonlinear descriptor systems". In: *Systems and Control Letters* 32.1, pp. 43–48. ISSN: 01676911. DOI: 10.1016/S0167-6911(97)00054-6.

In: An Introduction to the Extended ...
Editor: M. Holland

ISBN: 978-1-53618-875-2
© 2020 Nova Science Publishers, Inc.

Chapter 3

CONVERGENCE OF THE DISCRETE EXTENDED KALMAN FILTER IN NONLINEAR DETERMINISTIC SYSTEMS WITH NOISY OUTPUTS

Leonardo Esau Herrera, PhD and Jaime Herrera, PhD*
Mechanical and Aerospace Engineering Department,
NPS, Monterey, CA, US
Dirección de estudios de posgrado, CICESE,
Ensenada, B.C., Mexico

Abstract

The convergence of the extended Kalman filter in nonlinear discrete deterministic systems corrupted by white Gaussian noise in its outputs is presented. In the linear counterpart, the convergence was shown for the Kalman filter. To extend the results to the nonlinear frame and consequently cover a larger class of systems, useful tools from the linear counterpart are brought to support the present study. Some assumptions in the linearized structure of the nonlinear systems and its outputs are sufficient to demonstrate, via Lyapunov analysis, that the estimated state provided by the extended Kalman filter converges to that of the examined nonlinear systems. As a deterministic focus is considered, asymptotic convergence is achieved. The proposed contribution thus gives an insight into

*Corresponding Author's Email: leonardo.herrera@nps.edu.

the intrinsic convergence of the extended Kalman filter when operated in the stochastic frame for the class of systems and outputs considered. As a testbed, bifurcation parameter estimation is addressed in a three-dimensional chaotic system and its output to illustrate the convergence. Numerical results show the asymptotic convergence.

Keywords: extended Kalman filter, convergence, Lyapunov stability

1. Introduction

The Extended Kalman Filter (EKF) is a generalized version of the well-known Kalman Filter (KF). This version estimates the state vector of noisy nonlinear systems under noisy outputs; at the same time, the optimality property of the conventional Kalman filter is locally held. Despite local optimality, it has been useful from the Apollo Moon project to the present, see e.g., McGee and Schmidt (1985) and Grewal and Andrews (2010).

An essential property of the EKF that is not typically addressed in many types of researches is its convergence; this property guarantees the proper functioning of the filter, and its consideration would be crucial every time the EKF is employed. Some works dealing with the EKF convergence, although confined to the deterministic frame, can be seen in Boutayeb et al. (1997), Song and Grizzle (1992), Reif and Unbehauen (1999), and Guo and Zhu (2002), whereas works dealing with the stochastic convergence can be seen e.g., in La Scala et al. (1995), Reif et al. (1999), and Rapp and Nyman (2004).

As analyzing the convergence in the deterministic frame is more pleasant, and also provides an insight into the intrinsic behavior of the EKF when operated as stochastic, its study results of great interest. The former cited works analyze the convergence of the EKF in nonlinear discrete deterministic systems with discrete deterministic outputs; at the same time, they consider fictitious noises affecting the systems and its outputs, i.e., noises with zero values but arbitrary covariance matrices. These works are briefly detailed below.

In Boutayeb et al. (1997) two variant diagonal matrices are involved in the first-order approximations of the considered systems and outputs to recover an exact representation. Under these matrices is shown, in the convergence analysis, that the matrix R_k associated with the covariance of noise affecting the outputs plays an essential roll in enlarging the domain of attraction for the convergence. In Song and Grizzle (1992), bounded error covariance matrices of the EKF are sufficient to demonstrate local asymptotic convergence of the EKF; it is further shown that boundedness of the previous covariance matrices is satisfied with the observability property of the linearized structure of the considered nonlinear systems. In Reif and Unbehauen (1999), with a Lyapunov function and under a weight involved in the EKF is proved that the dynamics of the estimation error are locally exponentially stable; the prescribed weight defines the decay rate of the estimation error. In Guo and Zhu (2002), an improvement in the domain of convergence and the convergence rate of the EKF operated as an asymptotic observer is presented. On the one side, enlarging the domain of convergence is dependent on the covariances matrices Q_{k-1} and R_k associated respectively to noises in the considered systems and outputs. On the other side, increasing the convergence rate is achieved by decreasing the initial estimation error, a Neural Network is trained to keep the initial estimation error minimized.

Different from the previous literature, where the covariance matrices of the fictitious noises associated with the systems and outputs are considered, the present work contributes to analyzing the convergence when only the outputs are affected by fictitious noise. It means when the covariance matrix of the fictitious noise associated with the systems is zero, and the associated with the outputs is arbitrary. With this in mind, the intrinsic behavior of the EKF is thus analyzed in deterministic discrete systems with outputs corrupted by noise. These systems seem to be of paramount importance as they appear in many areas, e.g., system identification Zhang (2004), in training neural networks Williams (1992), and in processes represented by very accurate models.

Following the work of Zhang (2017), which is confined to the convergence of the KF for linear deterministic systems corrupted by noise in its linear outputs, the present work goes one step beyond by extending the results to the nonlinear case. The present work is organized as follows, section 2 states the problem to be solved, section 3 provides the convergence analysis that gives solution to the problem, section 4 defines a case of study, section 5 shows numerical results, and section 5.2 presents the conclusions.

2. PROBLEM STATEMENT

The problem consists in demonstrating that the EKF is asymptotically convergent in nonlinear discrete deterministic systems with outputs affected by fictitious noise. Proving the above convergence is of paramount importance as it provides an insight into the intrinsic behavior of the EKF when operated with stochastic outputs. Lyapunov stability theory is the tool to demonstrate the convergence, as it will be shown in the next section. To formally state the problem, the considered systems and its outputs, as well as the EKF, are first defined. The nonlinear systems considered are described by the deterministic difference equation

$$x_k = f(x_{k-1}), \tag{1}$$

where x_{k-1} represents the state vector and $f(x_{k-1})$ a nonlinear vectorial function that defines the dynamics of the systems. For (1), the outputs described by

$$y_k = h(x_k) + v_k, \tag{2}$$

are considered, where $h(x_k)$ is a function of the state vector and the random sequence v_k corresponds to uncertainties of White Gaussian Noise (WGN) normally distributed as $\mathcal{N}(0, R_k)$. R_k is the covariance matrix of v_k that allows tuning the EKF.

The EKF, due to Stanley F. Schmidt and his staff McGee and Schmidt (1985), estimates the state vector of (1) under WGN, i.e. of

$$x_k = f(x_{k-1}) + w_k, \tag{3}$$

where w_k are also uncertainties of WGN normally distributed as $\mathcal{N}(0, Q_k)$ with Q_k being a covariance matrix of w_k. To do so, the filter uses the outputs (2) and the knowledge of (3). This filter can be seen, *e.g.*, in Simon (2006) and Kim (2011). As in this work (1) is considered, the traditional EKF is reduced to

$$\hat{x}_k^p = f(\hat{x}_{k-1}^u), \tag{4}$$
$$P_k^p = A_{k-1} P_{k-1}^u A_{k-1}^T, \tag{5}$$
$$K_k = P_k^p C_k^T (C_k P_k^p C_k^T + R_k)^{-1}, \tag{6}$$
$$\hat{x}_k^u = \hat{x}_k^p + K_k (y_k - h(\hat{x}_k^p)), \tag{7}$$
$$P_k^u = (I - K_k C_k) P_k^p, \tag{8}$$

where the traditional equation $P_k^p = A_{k-1} P_{k-1}^u A_{k-1}^T + Q_{k-1}$ of the EKF has changed to (5). In the above relations, $\hat{x}_k^p$ is the estimated state of the true state x_k before taking y_k, P_k^p is the covariance matrix of the estimation error before taking y_k, K_k is the core of the EKF *i.e.* the Kalman gain, $\hat{x}_k^u$ describes the estimated state of x_k once y_k is present, and P_k^u is the covariance matrix of the estimation error once y_k is present as well, R_k is a symmetric and positive definite matrix that captures the variances of the fictitious uncertainties v_k in its diagonal, P_k^u and P_k^p are symmetric and positive definite matrices due to the assumptions that are going to be stated later. The superindexes p and u (p and u) are to indicate the prediction and update phases of the variables. On the one side, the predicted estimated state $\hat{x}_k^p$ as well as the predicted covariance matrix P_k^p are respectively computed as a function of the previous updated estimated state $\hat{x}_{k-1}^u$ and the previous updated covariance matrix P_{k-1}^u. On the other side, the updated estimated state $\hat{x}_k^u$ is computed as a function of the current predicted estimated $\hat{x}_k^p$ and the current measurement y_k, whereas the updated covariance matrix P_k^u is computed as a function of the current predicted covariance matrix P_k^p. The matrices

$$A_{k-1} = \left.\frac{\partial f(x_{k-1})}{\partial x_{k-1}}\right|_{\hat{x}_{k-1}^u}$$

and

$$C_k = \left.\frac{\partial h(x_k)}{\partial x_k}\right|_{\hat{x}_k^p}$$

come from Taylor's linear approximations of (1) and (2) around the nominal values $x_{k-1} = \hat{x}_{k-1}^u$, $x_k = \hat{x}_k^p$, and $v_k = 0$ (see, e.g., Simon (2006)). These approximations result as

$$x_k \approx A_{k-1}x_{k-1} + u_{k-1} \text{ and} \tag{9}$$

$$y_k \approx h(\hat{x}_k^p) + C_k x_k - C_k \hat{x}_k^p + v_k, \tag{10}$$

with $u_{k-1} = f(\hat{x}_{k-1}^u) - A_{k-1}\hat{x}_{k-1}^u$. It is appreciated that the EKF (4)-(8) results from the standard KF applied to estimate the state vector of (9) and the outputs

$$z_k = C_k x_k + v_k, \tag{11}$$

where $z_k = y_k - h(\hat{x}_k^p) + C_k \hat{x}_k^p$, (see Simon (2006)). The idealized initial conditions for the EKF are $\hat{x}_0^u = E[x_0]$ and $P_0^u = \text{var}[x_0]$, where E is the expectation operator, and var the variance.

Given the state x_k of (1) and the estimated $\hat{x}_k^u$ provided by (7), the problem consist in demonstrate that $\hat{x}_k^u \to x_k$ as $k \to \infty$ provided that $y_k = h(x_k)$. To deal with the problem, the following estimation error dynamics, resulting from the difference between x_k and $\hat{x}_k^u$, are defined

$$x_k - \hat{x}_k^u = f(x_{k-1}) - \hat{x}_k^p - K_k(y_k - h(\hat{x}_k^p)). \tag{12}$$

The next section provides sufficient conditions guaranteeing the asymptotic convergence to zero of (12), provided that $y_k = h(x_k)$.

3. Convergence Analysis

To prove the convergence, the powerful Lyapunov theory is brought to the present study. It is known that the qualitative behavior of dynamic systems can be deduced from the use of Lyapunov functions. To establish asymptotic convergence for the EKF deal in this work, a quadratic Lyapunov function, in terms of the estimation error, is proposed. It is thus deduced the qualitative behavior of asymptotic convergence for the EKF. The uniform observability and additional properties extracted from (9) and (11) are employed to prove the asymptotic convergence.

As a first step in the analysis, the internal dynamics of (12) ((12) with $y_k = h(x_k)$), are approximated to a linear dynamics. Through relations and (4), (9), and (10) applied to the internal dynamics of (12), the following linear dynamics are obtained

$$e_k = (I - K_k C_k) A_{k-1} e_{k-1}, \qquad (13)$$

where e_k is a state vector that locally represents the estimation error, $x_k - \hat{x}_k^u$, around $x_{k-1} = \hat{x}_{k-1}^u$ and $x_k = \hat{x}_k^p$. Due to the linear approximations (9) and (10), (13) only represents locally the estimation error dynamics.

Asymptotic stability for (13) is proved in terms of the Kalman Predictor (KP) error dynamics. According to Zhang (2017), the KP error dynamics are deduced from (13), where (13) coincides with the KF error dynamics. The differences between the KF and the KP can be found in Teixeira (2008). Following the analysis from Zhang (2017), the asymptotic convergence is then proved for the KP error dynamics and consequently established for (13). To deduce the KP error dynamics, let ζ_k be a sequence that is related with e_k through

$$e_k = (I - K_k C_k)\zeta_k, \qquad (14)$$

where the matrix $(I - K_k C_k)$, that is the same from (8), is invertible due to the positive definite property of P_k^u and P_k^p. By replacing (14) into (13), the following is obtained

$$(I - K_k C_k)\zeta_k = (I - K_k C_k) A_{k-1} (I - K_{k-1} C_{k-1})\zeta_{k-1}. \qquad (15)$$

Removing the factor $(I - K_k C_k)$ from both sides of the equality, the KP error dynamic is thus obtained as

$$\zeta_k = A_{k-1}(I - K_{k-1} C_{k-1})\zeta_{k-1}. \qquad (16)$$

This relation can also be rewritten as

$$\zeta_k = A_{k-1}\zeta_{k-1} - A_{k-1}K_{k-1}C_{k-1}\zeta_{k-1}. \tag{17}$$

It is notorious that the asymptotic convergence of (13) is guaranteed through the asymptotic convergence of (16) by means of relation (14). To state a theorem establishing the asymptotic convergence of the EKF considered, the following assumptions and lemmas inherited from Zhang (2017) are introduced.

Assumptions. *Relations (9) and (11) are defined with bounded real matrices A_{k-1} and C_k of appropriate dimension. For all $k \geq 0$, the square matrix A_{k-1} is invertible, such that $x_k = \Phi_{k,l} x_l$ -with $\Phi_{k,l}$ being a transition matrix- holds for all integers $k, l \geq 0$, including the case $k < l$. The matrix R_k is assumed to be a bounded real matrix that is also symmetric positive definite and with a strictly positive lower bound. The initial state x_0 follows the the normal Gaussian distribution $\mathcal{N}(\mu_0, P_0)$ with some mean vector μ_0 and a symmetric positive definite covariance matrix P_0. Additionally it is assumed that the pair $\{A_{k-1}, C_k\}$ is uniformly completely observable.*

Lemma 1. *Given relation (5), then*

$$(P^u_{k-1})^{-1} = A^T_{k-1}(P^p_k)^{-1}A_{k-1}. \tag{18}$$

Proof. *Inverting both sides of (5)*

$$(P^p_k)^{-1} = (A^T_{k-1})^{-1}(P^u_{k-1})^{-1}A^{-1}_{k-1}, \tag{19}$$

now multiplying A^T_{k-1} by the left, and A_{k-1} by the right of (19), lemma 1 is thus proved. □

Lemma 2. *Given relation (8), then*

$$P^u_{k-1}(P^p_{k-1})^{-1} = (I - K_{k-1}C_{k-1}). \tag{20}$$

Proof. *Multiplying $(P^p_k)^{-1}$ by the right of (8) results*

$$P^u_k(P^p_k)^{-1} = (I - K_k C_k), \tag{21}$$

now by moving the index one step back ($k = k-1$), lemma 2 is then proved. □

Lemma 3. *Given relation (6), then*

$$K_{k-1}^T(P_{k-1}^p)^{-1} = \alpha_{k-1}^{-1}C_{k-1}. \tag{22}$$

Proof. *Multiplying $(P_k^p)^{-1}$ by the left of (6)*

$$(P_k^p)^{-1}K_k = C_k^T \alpha_k^{-1} \tag{23}$$

with $\alpha_k = (C_k P_k^p C_k^T + R_k)$, now transposing both sides of (23) and moving again the index one step back

$$K_{k-1}^T[(P_{k-1}^p)^{-1}]^T = [\alpha_{k-1}^{-1}]^T C_{k-1}. \tag{24}$$

By the matrix property $[(\cdot)^T]^{-1} = [(\cdot)^{-1}]^T$ and by the symmetry of P_{k-1}^p and α_{k-1}, lemma 3 is then proved. □

Lemma 4. *Given the state transition matrix $\bar{\Phi}_{k,i}$ of (16), such that $\zeta_k = \bar{\Phi}_{k,i}\zeta_i$, and under the previously stated assumptions, there exist a constant $\rho > 0$ such that*

$$\rho I \leq \sum_{i=k-\beta}^{k} \bar{\Phi}_{i,k}^T \lambda_i \bar{\Phi}_{i,k} \tag{25}$$

for all $k \geq \beta$, and with the positive integer β, and $\lambda_i = C_i^T R_i^{-1} C_i$.

Proof. *This proof is given in the appendix of Zhang (2017). Lemma 4 of this work corresponds to lemma 1 of Zhang (2017).* □

Theorem. *Given the above assumptions and provided that $y_k = h(x_k)$, the EKF (4)-(8) asymptotically converges.*

Proof. *Since (13) locally describes the estimation error dynamics, its convergence to zero is proved through the KP error dynamics (16). As a consequence, the local convergence of the EKF is ensured. Let the discrete Lyapunov function*

$$V_k - V_{k-1} = \zeta_k^T(P_k^p)^{-1}\zeta_k - \zeta_{k-1}^T(P_{k-1}^p)^{-1}\zeta_{k-1}. \tag{26}$$

Replacing (16) in the above relation

$$V_k - V_{k-1} = \zeta_{k-1}^T \Pi \zeta_{k-1} - \zeta_{k-1}^T(P_{k-1}^p)^{-1}\zeta_{k-1}, \tag{27}$$

where

$$\Pi = (I - K_{k-1}C_{k-1})^T A_{k-1}^T (P_k^p)^{-1} A_{k-1}(I - K_{k-1}C_{k-1}). \tag{28}$$

By lemma 1

$$\Pi = (I - K_{k-1}C_{k-1})^T (P_{k-1}^u)^{-1} (I - K_{k-1}C_{k-1}), \tag{29}$$

now by lemma 2

$$\Pi = (P_{k-1}^p)^{-1} - C_{k-1}^T K_{k-1}^T (P_{k-1}^p)^{-1}, \tag{30}$$

and finally by lemma 3

$$\Pi = (P_{k-1}^p)^{-1} - C_{k-1}^T \alpha_{k-1}^{-1} C_{k-1}. \tag{31}$$

Relation (27) is then expressed as

$$V_k - V_{k-1} = -\zeta_{k-1}^T C_{k-1}^T \alpha_{k-1}^{-1} C_{k-1} \zeta_{k-1}. \tag{32}$$

Due to the previously stated assumptions, the matrix P_k^p computed from (5) is symmetric positive definite and upper bounded for all $k \geq 0$ (see corollary 1 of Zhang (2017)). Consequently $C_k P_k^p C_k^T$ is positive semidefinite and upper bounded by $\sigma_1 > 0$, where σ_1 is an upper bound of the largest eigenvalue of $C_k P_k^p C_k^T$ for all $k \geq 0$. As it was also assumed that the matrix R_k has a positive lower bound, then $\sigma_2 > 0$ is a lower bound of the smallest eigenvalue of R_k for all $k \geq 0$. From this follows

$$\frac{\sigma_1}{\sigma_2} R_k + R_k = \sigma R_k \geq C_k P_k^p C_k^T + R_k \tag{33}$$

whit $\sigma = \frac{\sigma_1}{\sigma_2} + 1$. Under this inequality, the Lyapunov function (32) satisfy

$$V_k - V_{k-1} \leq -\frac{1}{\sigma} \zeta_{k-1}^T \lambda_{k-1} \zeta_{k-1} \leq 0, \tag{34}$$

with $\lambda_{k-1} = C_{k-1}^T R_{k-1}^{-1} C_{k-1}$. So far Lyapunov stability has been proved for (16); however, it still remains to prove asymptotic stability. By propagating inequality (34) the following is obtained

$$V_{k+1} - V_{k-\beta} \leq -\frac{1}{\sigma} \sum_{i=k-\beta}^{k} \zeta_i^T \lambda_i \zeta_i \tag{35}$$

$$= -\frac{1}{\sigma} \zeta_k^T \sum_{i=k-\beta}^{k} \bar{\Phi}_{i,k}^T \lambda_i \bar{\Phi}_{i,k} \zeta_k, \tag{36}$$

where β is the positive integer of lemma 4, and $\bar{\Phi}_{i,k}$ the state transition matrix of (16) such that $\zeta_i = \bar{\Phi}_{i,k} \zeta_k$. By lemma 4

$$V_{k+1} - V_{k-\beta} < -\frac{\rho}{\sigma} \zeta_k^T \zeta_k = -\frac{\rho}{\sigma} ||\zeta_k||^2, \tag{37}$$

then it follows that (16) is asymptotically stable in the Lyapunov sense. Theorem 1 is thus proved. □

4. Chaotic System as a Case of Study

A three-dimensional discrete system that, under some parametric values, exhibits chaos is considered in this section as a case of study to illustrates the convergence. This system is due to Méndez-Ramírez et al. (2017) and its continuous version is as follows

$$\begin{aligned} \dot{x}_1 &= -ax_1 - bx_2 x_3, \\ \dot{x}_2 &= -x_1 + cx_2, \\ \dot{x}_3 &= d - (x_2)^2 - x_3, \end{aligned}$$

where a, b, c, and d are scalars and positive parameters, and x_1, x_2, and x_3 its states. By the Euler's method, its discrete version results in the following three dimensional system

$$\begin{aligned} x_k^1 &= x_{k-1}^1 + \tau(-ax_{k-1}^1 - bx_{k-1}^2 x_{k-1}^3), & (38) \\ x_k^2 &= x_{k-1}^2 + \tau(-x_{k-1}^1 + cx_{k-1}^2), & (39) \\ x_k^3 &= x_{k-1}^3 + \tau(d - (x_{k-1}^2)^2 - x_{k-1}^3), & (40) \end{aligned}$$

where a, b, c, and d are still the same parameters as in its continuous version, whereas τ is a new parameter that comes from the discretization, it represents the step size between the sample $k-1$ and k. The bifurcation parameters, that define the existence or absence of chaos, in this version are b, d, and τ, whereas the states are x_{k-1}^1, x_{k-1}^2, and x_{k-1}^3. The Euler's method has been considered to be consistent with Méndez-Ramírez et al. (2017), but any other can be used, e.g., Runge-Kutta, Leapfrog, etc.. A simulation for the discrete system (38)-(40) under its chaotic operation mode is shown in Fig. 1. According with the bifurcation diagram from Méndez-Ramírez et al. (2017), this operation mode is, *e.g.*, reached by setting the bifurcation parameters as $b = 2, d = 4$, and $\tau = 0.085$, whereas setting the remaining parameters as $a = 2$ and $c = 0.5$, and the initial conditions as $x_0^1 = x_0^2 = x_0^3 = 1$. The simulation produced in Fig. 1 was run under these last parameters and initial conditions, and for $k = 1, \cdots, n$, with $n = 2000$.

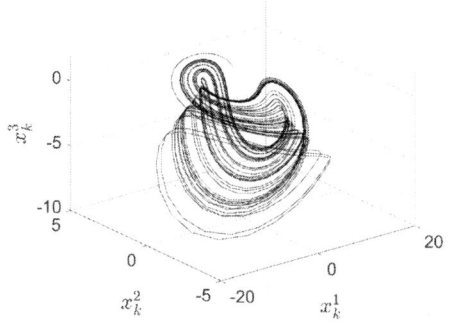

Figure 1. Chaotic behavior of the discrete system.

To estimate the states and the bifurcation parameters of (38)-(40), it is augmented with the dynamic of the bifurcation parameters. This augmentation is to represent the bifurcation parameters as states of the system and therefore being estimated via EKF. The augmented system is described by (1), specified with

$$\boldsymbol{x}_{k-1}^T = [x_{k-1}^1 \ x_{k-1}^2 \ x_{k-1}^3 \ b_{k-1} \ d_{k-1} \ \tau_{k-1}],$$

and

$$\boldsymbol{f}(\boldsymbol{x}_{k-1}) = \begin{bmatrix} x_{k-1}^1 + \tau_{k-1}(-ax_{k-1}^1 - b_{k-1}x_{k-1}^2 x_{k-1}^3) \\ x_{k-1}^2 + \tau_{k-1}(-x_{k-1}^1 + cx_{k-1}^2) \\ x_{k-1}^3 + \tau_{k-1}(d_{k-1} - (x_{k-1}^2)^2 - x_{k-1}^3) \\ b_{k-1} \\ d_{k-1} \\ \tau_{k-1} \end{bmatrix}.$$

For this system, the output is described by (2), and is specified with $\boldsymbol{h}(\boldsymbol{x}_k)^T = [x_k^1 \ x_k^2 \ x_k^3]$ and $\boldsymbol{v}_k^T = [v_k^1 \ v_k^2 \ v_k^3]$.

5. NUMERICAL RESULTS

Numerical results illustrating the convergence of the EKF are presented in this section. The results are obtained under the parameters shown in table 1, that contains the parametric values of (1)-(2) and the filter (4)-(8). The simulation to obtain the results was run with $n = 1000$ samples, and the index $k = 1, \cdots, n$ was changed by the time variable $t = \tau k$, where τ stands for the time step. Although the work was addressed to the fictitious uncertainties case ($v_t = 0$), the nonfictitious uncertainties case ($v_t \neq 0$) is also simulated for illustrative purposes. In the nonfictitious uncertainties case v_t is WGN normally distributed as it is shown in table 1. Simulation results are then shown for both cases.

Table 1. Augmented chaotic system, its output, and EKF parameters

Parameter	Value
a	2
b	2
c	0.5
d	4
τ	0.085
v_t	$\mathcal{N} \sim (0, R_t)$
R_t	$0.08 I_{3\times 3}$
x_0	$[1\ 1\ 1]^T$
$\hat{x}_0^u$	$[1.1\ 1.1\ 1.1\ 1.9\ 3.9\ 0.095]^T$
P_0^u	$0.01 I_{6\times 6}$

5.1. Fictitious Uncertainties Case: $v_t = 0$

Figures 2a, 2b, and 3a show the estimations of the bifurcation parameters b, d, and τ, respectively. In these figures, $\hat{b}_t^u, \hat{d}_t^u$, and $\hat{\tau}_t^u$ converge asymptotically to the true parametric values. The estimation error variances associated with $\hat{b}_t^u, \hat{d}_t^u$, and $\hat{\tau}_t^u$ are depicted in table 2. These variances are obtained from a) the estimation error dynamics (12) and b) the local estimation error dynamics (13). From both a) and b), the estimation error variance that shows less magnitude is the associated with $\hat{\tau}_t^u$, after this follows the associated with $\hat{b}_t^u$, and then the associated with $\hat{d}_t^u$. This can be reflected through the transient deviations of the estimations in Fig. 2a, 2b, and 3a. Figure 3b presents the estimation error variances extracted from the diagonal of P_t^u and associated with $\hat{b}_t^u$, $\hat{d}_t^u$, and $\hat{\tau}_t^u$. These variances also indicate that the estimated error with less variance is the associated with $\hat{\tau}_t^u$, then the associated with $\hat{b}_t^u$, and finally the associated with $\hat{d}_t^u$. Figure 4a shows the convergence to zero of the norm 2 of the estimation error that comes from (12). Figure 4b shows the convergence to zero of the norm 2 of the estimation error that comes from the approximated dynamics (13). Figures 4a and 4b are pretty similar due that (13) represents well enough (12) during the estimation process simulated in this section. Figure 5 shows the phase portrait produced by the output y_t of the chaotic system (38)-(40), under this output the estimation was carried out.

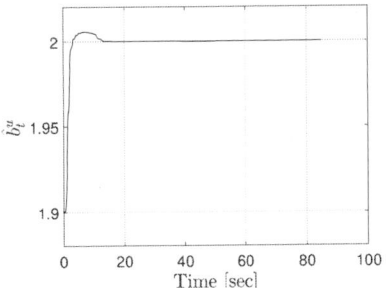

(a) Estimation of the parameter b. The estimation is initialized in $\hat{b}_0^u = 1.9$ and converges to the true parametric value $b = 2$ approximately after 20 seconds.

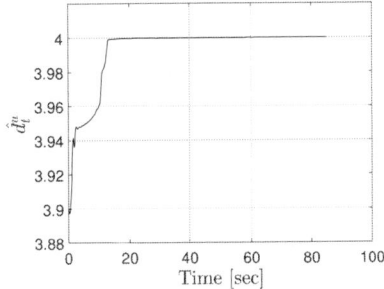

(b) Estimation of the parameter d. The estimation is initialized in $\hat{d}_0^u = 3.9$ and converges to the true parametric value $d = 4$ approximately after 20 seconds.

Figure 2.

Table 2. Estimation error variances for the fictitious uncertainties case

	$E[(b-\hat{b}_t^u)^2]$	$E[(d-\hat{d}_t^u)^2]$	$E[(\tau-\hat{\tau}_t^u)^2]$	$E[(e_t^4)^2]$	$E[(e_t^5)^2]$	$E[(e_t^6)^2]$
$v_t = 0$	1.52×10^{-4}	4.24×10^{-4}	2.63×10^{-7}	1.51×10^{-4}	3.41×10^{-4}	2.65×10^{-7}

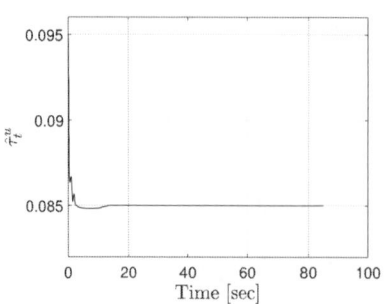
(a) Estimation of the parameter τ. The estimation is initialized in $\hat{\tau}_0^u = 0.095$ and converges to the true parametric value $\tau = 0.085$ approximately after 20 seconds.

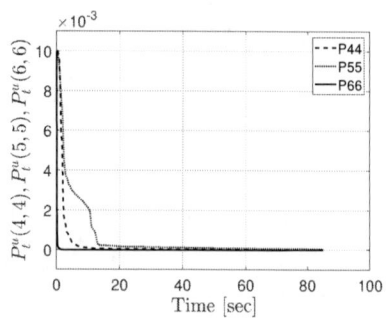
(b) Variances of the estimation errors extracted from P_t^u. $P_t^u(4,4), P_t^u(5,5)$, and $P_t^u(6,6)$ are the variances of the estimation errors associated with $\hat{b}_t^u, \hat{d}_t^u$, and $\hat{\tau}_t^u$ respectively.

Figure 3.

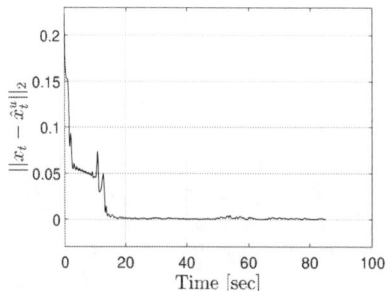
(a) Norm 2 of the estimation error. This error is obtained from (12).

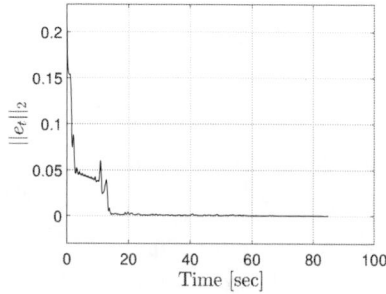
(b) Norm 2 of the estimation error. This error is obtained from (13).

Figure 4.

5.2. Nonfictitious Uncertainties Case: $v_t \neq 0$

Figures 6a, 6b, and 7a show the estimations of the bifurcation parameters b, d, and τ, respectively. In these figures, the estimated parameters $\hat{b}_t^u, \hat{d}_t^u$, and $\hat{\tau}_t^u$ don't converge asymptotically to the true parametric values due to the nonfictitious uncertainties v_t; however, these remain in the closeness of the parametric true values. Similar to the fictitious uncertainties case, the estimation error vari-

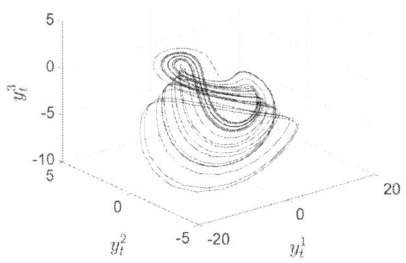

Figure 5. Phase portrait produced by the output y_t of the chaotic system (38)-(40).

ances associated with $\hat{b}^u_t$, $\hat{d}^u_t$, and $\hat{\tau}^u_t$ are depicted in table 3. These variances are obtained from a) the estimation error dynamics corresponding to (12) but deduced under $v_t \neq 0$ and b) the local estimation error dynamics corresponding to (13), but also deduced under $v_t \neq 0$. From both a) and b), the estimation error variance that shows less magnitude is the associated with $\hat{\tau}^u_t$, after this follows the associated with $\hat{b}^u_t$, and then the associated with $\hat{d}^u_t$. This also can be reflected through the transient deviations of the estimations in Fig. 6a, 6b, and 7a. Figure 7b presents the estimation error variances extracted from the diagonal of P^u_t and associated with $\hat{b}^u_t$, $\hat{d}^u_t$, and $\hat{\tau}^u_t$. These variances also indicate that the estimation error with less variance is the associated with $\hat{\tau}^u_t$, then the associated with $\hat{b}^u_t$, and finally the associated with $\hat{d}^u_t$. Figure 8a shows the norm 2 of the estimation error that comes from the dynamics corresponding to (12), but deduced under $v_t \neq 0$. Figure 8b presents the norm 2 of the estimation error that comes from the approximated dynamic corresponding to (13) but deduced accordingly, under $v_t \neq 0$. Figures 8a and 8b are pretty similar due that (13) represents well enough (12) during the estimation process. Finally, Figure 9 shows the phase portrait produced by the output y_t of the chaotic system (38)-(40).

Table 3. Estimation error variances for the nonfictitious uncertainties case

	$E[(b-\hat{b}^u_t)^2]$	$E[(d-\hat{d}^u_t)^2]$	$E[(\tau-\hat{\tau}^u_t)^2]$	$E[(e^4_t)^2]$	$E[(e^5_t)^2]$	$E[(e^6_t)^2]$
$v_t \neq 0$	3.53×10^{-4}	2.40×10^{-3}	8.25×10^{-6}	2.87×10^{-4}	9.07×10^{-4}	8.85×10^{-6}

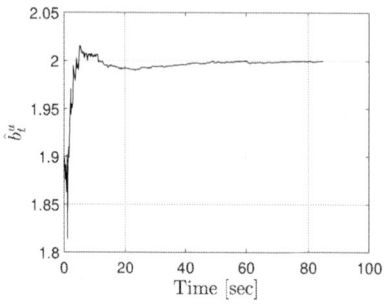
(a) Estimation of the parameter b. The estimation is initialized in $\hat{b}_0^u = 1.9$ and remains in the closeness of the true parametric value $b = 2$.

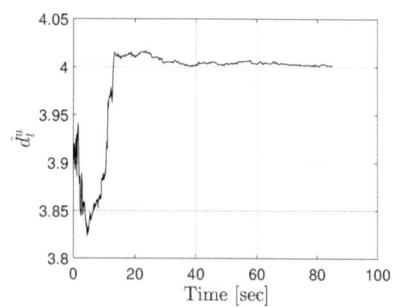
(b) Estimation of the parameter d. The estimation is initialized in $\hat{d}_0^u = 3.9$ and remains in the closeness of the true parametric value $d = 4$.

Figure 6.

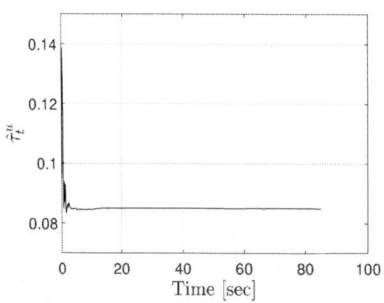
(a) Estimation of the parameter τ. The estimation is initialized in $\hat{\tau}_0^u = 0.095$ and remains in the closeness of the true parametric value $\tau = 0.085$.

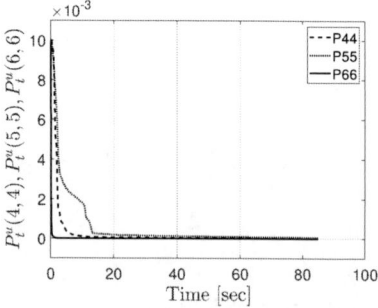
(b) Variances of the estimation errors extracted from P_t^u. $P_t^u(4,4), P_t^u(5,5)$, and $P_t^u(6,6)$ are the variances of the estimation errors associated with $\hat{b}_t^u, \hat{d}_t^u$, and $\hat{\tau}_t^u$ respectively.

Figure 7.

CONCLUSION

Asymptotic convergence of the EKF was proved for nonlinear discrete deterministic systems corrupted by fictitious white Gaussian noise in their outputs. Some assumptions were sufficient to demonstrate, via Lyapunov analysis, the

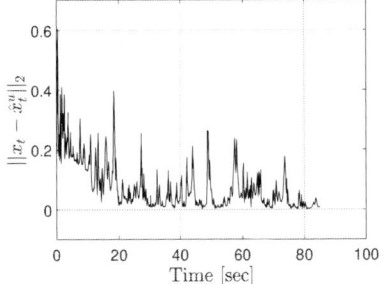
(a) Norm 2 of the estimation error. This error is obtained from (12).

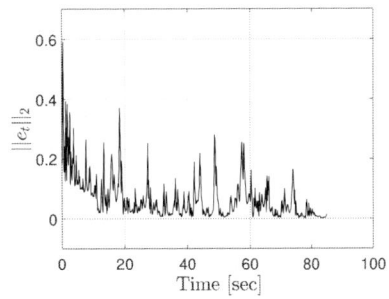
(b) Norm 2 of the estimation error. This error is obtained from (13).

Figure 8.

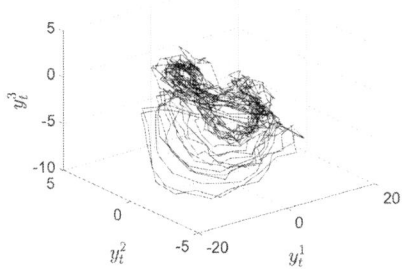

Figure 9. Phase portrait produced by the output y_t of the chaotic system (38)-(40).

convergence. Under the proved convergence, an insight into the intrinsic behavior of the EKF when operated in a stochastic frame can hence be deduced. As a testbed, asymptotic estimation of bifurcation parameters in a three-dimensional discrete system and its output was performed. Although a deterministic analysis was considered, numerical results were also shown for the stochastic. In the deterministic frame, asymptotic convergence was shown, whereas, in the stochastic, the estimated evolved around their references. Although the sufficient assumptions were not verified for the testbed, the numerical results confirm the convergence. Thus, it is probable that the conditions are not only sufficient but also necessary. This attestation is left for future investigation.

REFERENCES

Boutayeb, M., H. Rafaralahy, and M. Darouach (1997). Convergence analysis of the extended Kalman filter used as an observer for nonlinear deterministic discrete-time systems. *IEEE transactions on automatic control 42*(4), 581–586.

Grewal, M. S. and A. P. Andrews (2010). Applications of Kalman filtering in aerospace 1960 to the present [historical perspectives]. *IEEE Control Systems Magazine 30*(3), 69–78.

Guo, L. and Q. Zhu (2002). A fast convergent extended Kalman observer for nonlinear discrete-time systems. *International Journal of Systems Science 33*(13), 1051–1058.

Herrera, L. and F. Jurado (2018). Asymptotic Identification of Bifurcation Parameters in a Three-Dimensional Discrete System with Chaotic Behavior. In *2018 IEEE Latin American Conference on Computational Intelligence (LA-CCI)*, pp. 1–5. IEEE.

Kim, P. (2011). *Kalman filter for beginners: with MATLAB examples*. CreateSpace.

La Scala, B. F., R. R. Bitmead, and M. R. James (1995). Conditions for stability of the extended Kalman filter and their application to the frequency tracking problem. *Mathematics of Control, Signals and Systems 8*(1), 1–26.

McGee, L. A. and S. F. Schmidt (1985). Discovery of the Kalman Filter as a Practical Tool for Aerospace and Industry. In *National Aeronautics and Space Administration, Ames Research*.

Méndez-Ramírez, R., A. Arellano-Delgado, C. Cruz-Hernández, and R. Martínez-Clark (2017). A new simple chaotic lorenz-type system and its digital realization using a tft touch-screen display embedded system. *Complexity 2017*.

Rapp, K. and P.-O. Nyman (2004). Stability properties of the discrete-time extended Kalman filter. *IFAC Proceedings Volumes 37*(13), 1377–1382.

Reif, K., S. Gunther, E. Yaz, and R. Unbehauen (1999). Stochastic stability of the discrete-time extended Kalman filter. *IEEE Transactions on Automatic control 44*(4), 714–728.

Reif, K. and R. Unbehauen (1999). The extended Kalman filter as an exponential observer for nonlinear systems. *IEEE Transactions on Signal processing 47*(8), 2324–2328.

Simon, D. (2006). *Optimal state estimation: Kalman, H infinity, and nonlinear approaches*. John Wiley & Sons.

Song, Y. and J. W. Grizzle (1992). The extended Kalman filter as a local asymptotic observer for nonlinear discrete-time systems. In *1992 American control conference*, pp. 3365–3369. IEEE.

Teixeira, B. (2008). Kalman filters [ask the experts]. *IEEE Control Systems 28*(2), 16–18.

Williams, R. J. (1992). Training recurrent networks using the extended Kalman filter. In *[Proceedings 1992] IJCNN International Joint Conference on Neural Networks*, Volume 4, pp. 241–246. IEEE.

Zhang, Q. (2004). Nonlinear system identification with output error model through stabilized simulation. *IFAC Proceedings Volumes 37*(13), 501–506.

Zhang, Q. (2017). On stability of the Kalman filter for discrete time output error systems. *Systems & Control Letters 107*, 84–91.

INDEX

A

azimuth, 2, 3, 4, 5, 7, 11, 12, 13, 14, 16, 24, 25, 32, 33, 35
azimuth angle, 13

B

bounds, 68

C

convergence, vii, viii, 4, 11, 35, 36, 83, 84, 85, 87, 88, 89, 90, 91, 92, 93, 94, 95, 97, 98, 99, 100, 101
covariance estimation, 10, 15
covariance prediction, 10, 15

D

differential equations, 49, 72
direct measure, 21
discretization, 92
distribution, 49, 58, 59, 89
dynamic systems, 79, 88

E

Earth-Centered-Earth-Fixed, 12
Earth-Centered-Inertial (ECI) coordinate frame, 6, 11, 12, 17, 18, 20, 35
East-North (UEN) coordinate system, 7, 11, 12, 17, 18, 20, 35
elevation, vii, 2, 3, 4, 5, 7, 11, 12, 13, 14, 16, 24, 25, 32, 33, 35
elevation angle, 4, 12, 13, 16
extended Kalman filter (EKF), vii, viii, 1, 2, 3, 4, 5, 7, 9, 11, 13, 14, 15, 16, 17, 18, 19, 21, 23, 25, 27, 29, 31, 33, 35, 36, 37, 39, 40, 41, 42, 55, 56, 57, 59, 77, 79, 83, 84, 85, 86, 87, 88, 89, 90, 91, 93, 94, 95, 97, 98, 99, 100, 101
extrapolation error, 37

F

fault detection, vii, viii, 44, 51, 52, 53, 54, 60, 61, 63, 65, 66, 67, 70, 74, 76, 79, 80
fault diagnosis, viii, 43, 45, 48, 49, 66, 69, 74, 76
filter-gain, 37
filters, vii, viii, 2, 3, 4, 5, 20, 21, 35, 41, 44, 53, 54, 59, 61, 66, 67, 69, 70, 74, 101

G

Gaussian white noise, 6, 56
Gaussian-mixture Kalman filters, 4
geostationary satellite, vii, 2, 3, 5, 16, 35, 42
geosynchronous orbits, 4
GPS measurement, 40
GPS receivers, 2

gravitational field, 3
ground station, 2, 12, 13

I

indirect measurements, 2, 28, 35, 36
initial state, 20, 54, 59, 89
innovation, 9, 14, 30, 67
isolation, viii, 44, 65, 66, 67, 74, 76, 77, 79, 80
iteration, 59

J

Jacobian/Jakobian, 8, 10, 11, 19, 30, 35, 39

K

Kalman filter, 3, 37, 38, 45, 47, 49, 51, 53, 55, 57, 59, 61, 63, 65, 67, 69, 71, 73, 75, 77, 79, 81, 84, 100
Kalman filter(ing), 3, 5, 12, 37, 38, 45, 47, 49, 51, 53, 55, 57, 59, 61, 63, 65, 67, 69, 71, 73, 75, 77, 78, 79, 80, 81, 84, 100
Kalman gain, 9, 15, 60, 86
Kepler constant, 6
Kepler equations, 5
Kronecker symbol, 8, 37

L

least squares estimation, 2
linear function, 13
linear systems, 48
liquid phase, 73
Lyapunov function, 85, 88, 90, 91
Lyapunov stability, 84, 85, 91

M

magnetometers, 2
matrix, 7, 9, 10, 11, 13, 14, 16, 19, 20, 21, 30, 31, 35, 36, 42, 49, 55, 56, 57, 60, 68, 73, 84, 86, 87, 88, 89, 90, 91, 92
maximum likelihood estimate, 54, 55

measurements, vii, 1, 2, 3, 4, 5, 8, 11, 12, 14, 16, 21, 22, 23, 24, 25, 26, 27, 31, 32, 33, 34, 35, 36, 40, 41, 42, 56, 57, 62, 63

N

navigation system, 3
neural networks, 52, 85
nonlinear systems, viii, 36, 42, 83, 85, 86, 101
normalized average errors, 28

O

one-stage prediction, 37
orbit determination, vii, 2, 3, 4, 5, 7, 9, 11, 12, 13, 15, 17, 19, 21, 23, 25, 27, 29, 31, 33, 35, 37, 39, 41, 42
orbit determination Kalman filter, 4

P

predicted observation, 8, 9

R

range, 1, 3, 4, 6, 11, 12, 13, 14, 16, 17, 22, 23, 24, 25, 32, 33, 35, 36, 46
rate of change, 61, 72
root mean squares, 27, 30

S

sensor(s), 2, 3, 4, 44, 46, 51, 66, 67, 70, 76, 80
signals, 3, 46
single station antenna, vii, 1, 2, 5, 11, 12, 35, 40, 41, 42
Space Surveillance Network (SSN), 4
spacecraft, vii, 1, 2, 3, 5, 6, 7, 11, 12, 13, 14, 16, 17, 20, 35, 40, 41, 42
spacecraft localization, 3, 40
stability, 44, 69, 84, 85, 88, 91, 100, 101
standard deviation, 16
star tracker sensor, 3
state estimation, 9, 14, 41, 53, 54, 59, 63, 68, 74, 75, 77, 78, 79, 101
state prediction, 8, 9, 14

states, 17, 21, 53, 54, 55, 56, 57, 58, 60, 69, 72, 85, 92
system noise covariance matrix, 21

T

techniques, 4, 52, 79
technology, 40, 41
transesterification, 72
transformation(s), 4, 12, 18, 20, 40, 57, 59
transformation matrix, 18

U

unscented transformation, 4, 40, 57

V

variables, 2, 5, 17, 20, 56, 68, 87
vector, 7, 8, 9, 14, 16, 17, 18, 19, 20, 36, 41, 49, 51, 52, 68, 69, 84, 86, 87, 88, 89
velocity, vii, 1, 2, 3, 4, 5, 11, 14, 16, 17, 21, 23, 25, 26, 27, 28, 29, 30, 31, 32, 33, 34, 35, 41